Model-Based Optimization for Petroleum Refinery Configuration Design

Model-Based Optimization for Petroleum Refinery Configuration Design

Cheng Seong Khor

WILEY-VCH

Author

Dr. Cheng Seong Khor
PETRONAS Digital (Data Science)
50088 Kuala Lumpur
Malaysia

Library of Congress Card No.: applied for

British Library Cataloguing-in-Publication Data
A catalogue record for this book is available from the British Library.

Bibliographic information published by the Deutsche Nationalbibliothek
The Deutsche Nationalbibliothek lists this publication in the Deutsche Nationalbibliografie; detailed bibliographic data are available on the Internet at <http://dnb.d-nb.de>.

Print ISBN: 978-3-527-34741-4
ePDF ISBN: 978-3-527-82609-4
ePub ISBN: 978-3-527-82608-7
oBook ISBN: 978-3-527-82610-0

Typesetting Straive, Chennai, India

Printed and bound by CPI Group (UK) Ltd, Croydon, CR0 4YY

C9783527347414_090124

Contents

1

Introduction to Optimization Modeling for Petroleum Refineries

1.1 Background

The topological optimization problem for determining an optimal configuration of a petroleum refinery can be addressed by a logic-based modeling approach within a mixed-integer superstructure optimization framework. The focus lies in investigating and advancing existing optimization approaches and strategies of employing logical constraints to conceptual process synthesis and design problems within the framework of conventional mixed-integer linear programming (MILP) (Nemhauser and Wolsey 1988) and alternate generalized disjunctive programming (GDP) (Grossmann and Trespalacios 2013). This work attempts to address the following considerations:

- How the formulation of design specifications in a synthesis problem can be accomplished using logical constraints in a mixed-logical-and-integer optimization model to enrich the problem representation by way of incorporating past design experience, engineering knowledge, and heuristics;
- How structural specifications on the interconnectivity relationships by space (states) and function (tasks) should be properly formulated using logical constraints within a mixed-integer optimization model.

The resulting modeling technique is illustrated on a numerical example, which is based on a case study involving alternative processing routes of naphtha in a refinery.

Process synthesis or conceptual process design is concerned with the identification of the best flowsheet structure to perform a given task. The following variants are mainly available in the literature to address this class of problem: (1) the heuristics method, notably the hierarchical decomposition of design decisions procedure; (2) the technique based on thermodynamic targets and physical insights as exemplified by pinch analysis; and (3) the algorithmic approach that utilizes optimization based on the construction of a superstructure that seeks to represent all feasible process flowsheets (Seider et al. 2009).

The intricate complexities associated with process synthesis problem in general and the refinery design problem in specific necessitates the development and implementation of a systematic and automated approach that efficiently and rigorously integrates the elaborate interactions involving the design decision variables.

Model-Based Optimization for Petroleum Refinery Configuration Design, First Edition. Cheng Seong Khor.

This work aims to extend the superstructure optimization-based approach of using logical constraints (Raman and Grossmann 1991, 1992, 1993a,b) within a MILP to incorporate qualitative design knowledge based on engineering experience and heuristics in modeling the major process flows in a refinery. These constraints adopt discrete integer decision variables of the binary 0–1 type to model the existence of a refinery process unit and the associated stream piping interconnections (which are effectively pipelines) in a network structure, in which a value of one for a 0–1 variable designates that a unit is present in the optimal structure while the converse is true for a value of zero.

Our work serves to further substantiate that the use of 0–1 decision variables offers a more natural and powerful modeling approach compared to the conventional linear programming technique that employs only continuous decision variables. It also affords the convenience of representing fixed-cost charges in the objective function formulation. A variation in the use of integer variables in optimization model formulations has been widely reported (Williams 1999).

Optimization is the core objective of chemical process design as exemplified through the synthesis of petroleum refinery configurations (Khor and Varvarezos 2017). Selecting the best among a set of possible solutions requires good engineering judgment to critically analyze the process with respect to the desired performance objectives. It is crucial to identify and strike a balance between the competing objectives of realizing the largest production, the greatest profit, the minimum cost, the least energy usage, and so on. This ensures improved plant performance through improved yields of valuable products, higher processing rates, longer time between shutdowns, and reduced maintenance costs. In order to find the best solution within the given constraints and flexibilities, a trade-off usually exists between capital and operating costs.

Although the design stage only takes up about 2% or 3% of a project expenditure, decisions made during this phase have an immense impact on plant economic performance because approximately 80% of the capital and operating expenses of the final plant are fixed during the design stage (Biegler et al. 1997). Hence, the necessity of developing systematic methods in chemical process design has led to two major strategies for process synthesis in determining an optimal configuration of a flowsheet and its operating condition.

In the first strategy, the problem can be solved in a sequential form involving decomposition, fixing some elements in the flowsheet, and then using heuristic rules to determine changes in the flowsheet that may lead to an improved solution. An example of such a strategy is the sequential hierarchical decomposition strategy by Douglas (1985, 1988). However, the sequential nature of the decisions and the heuristic rules that are used can lead to suboptimal designs. Douglas claims that only 1% of all designs are ever implemented in practice and hence this screening procedure avoids meticulous evaluation of most alternatives. It is not possible to rigorously produce an optimal design because the sequential nature of flowsheet synthesis cannot take all interactions among the design variables into consideration. Furthermore, the exponential number of possible topologies coupled with the multitude of process technology options decrease the chances of realizing the best design.

The second strategy that can be applied to solve a process synthesis problem is based on simultaneous optimization using mathematical programming (Grossmann 1996). This strategy requires the postulation of a superstructure, which includes a set of equipment that are potentially selected in the final flowsheet and their interconnections. The equations pertaining to the equipment and their interconnectivity in addition to the operating condition constraints are formulated in an optimization model with an objective function that typically minimizes cost or maximizes profit. In particular, such a formulation requires discrete variables to represent the choices of equipment besides continuous variables on the process parameters (e.g. flow rates) with which the model becomes a mixed-integer linear or nonlinear program (MILP or MINLP). In this regard, Grossmann (1996) states that an advantage of mathematical programming strategy is that they can perform simultaneous optimization of the configuration (as described by the discrete decisions) and operating conditions (as described by the continuous decisions).

Designing a petroleum refinery configuration is challenging and complex. Many factors such as design specifications and structural specifications have to be considered and incorporated at the conceptual design stage to arrive at an optimum configuration of the refinery flowsheet (Khor et al. 2011). Hierarchical decomposition uses heuristics, shortcut design procedures, and engineering experience to develop an initial base case, but doing so is possibly time-consuming, whereas the result may not necessarily guarantee an optimal solution. Thus, developing or adopting an automated systematic procedure in the refinery configuration design endeavor can significantly improve the decision-making process. The task can be achieved via optimization or mathematical programming approach by representing the problem through a superstructure and formulating the corresponding optimization model, which is solved to obtain an optimal configuration based on inputs of crude oils to be processed and final products to meet market demands while complying with the requisite constraints.

Figure 1.1 shows the rapidly rising downstream capital cost index from 2005 to 2019. Thus, an automated approach that can guarantee an optimal refinery design is increasingly important and sought after in the face of increased capital costs, higher energy costs, and depleting resources. At the same time, heightened fuel consumption leads to raised demand for petroleum products despite tight supplies with the consequential need to construct new grassroots petroleum refineries.

With increasingly stricter environmental regulations and emphasis on clean fuels, new refineries need to adhere to narrower operating margins and more stringent product specifications. This situation adds to the degree of complexity in designing refineries, which at present is already time-consuming with the intricacies of the interplay among the various factors including public opinions and permitting processes. All these considerations give rise to an exponential number of possible refinery topologies or configurations that can adequately meet current economic, operating, and environmental requirements.

We consider the following superstructure optimization problem for a refinery topology design. Given the following data: fixed production amounts of desired products, available process units and ranges of their capacities, and cost of crude oil

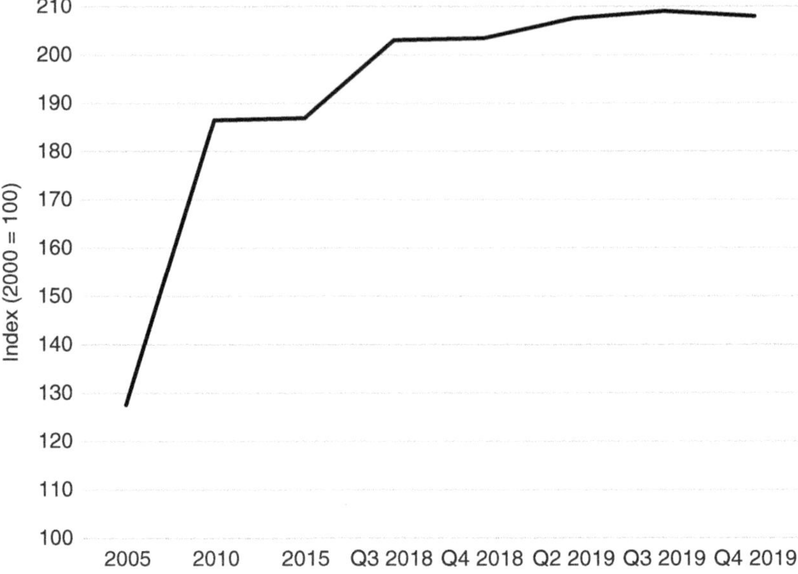

Figure 1.1 Downstream capital cost index. Source: Data taken from Oil & Gas Journal (Editorial) (2019).

and process units, we wish to determine an optimal configuration in terms of the unit selection and sequencing as well as the operating levels.

This work aims to present a primer on superstructure optimization by emphasizing the following aspects:

- developing a superstructure representation for a refinery network topology with a suitable level of detail;
- formulating models based on the superstructure representation by adopting two mixed-integer optimization frameworks: MILP and GDP, which incorporate both continuous and discrete decisions;
- solving the models using standard commercial off-the-shelf solvers enhanced with tailored solution strategies;
- analyzing and interpreting the model solution in terms of practical real-world applications.

A high-level view of the modeling approach adopted is shown in Figure 1.2.

1.2 Overview of Refining Processes

Petroleum products are made from crude oil. There are many types of crude oil from many different sources around the world. The selection of the right crude oil is a key part of the refining process. The decision as to what crude oil or combination of crude oils to process depends on many factors including quality, availability,

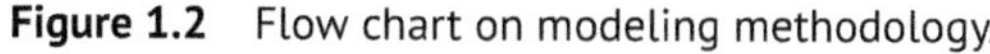

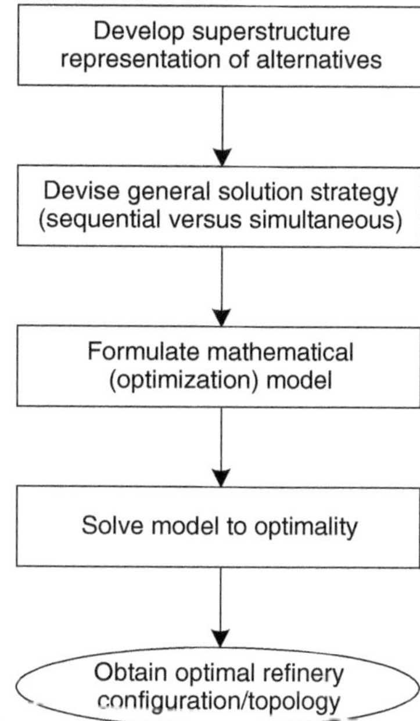

Figure 1.2 Flow chart on modeling methodology.

volume, and price. Table 1.1 briefly describes functions of several major refinery processes (Figure 1.3).

1.2.1 Atmospheric Crude Oil Distillation

The first stage of crude oil processing involves distillation or fractionation. The crude oil is distilled into fractions according to boiling point to yield light-end hydrocarbons (C_1–C_4), light naphtha, heavy naphtha, kerosene, diesel, and atmospheric residual. Some of these broad cuts can be marketed directly, while others require further processing in downstream units. Increased efficiency and reduced costs are achieved if the crude oil is fractionated at essentially atmospheric pressure followed by residue or bottoms fractionation using vacuum distillation (Figure 1.4).

Naphtha is a complex mixture of paraffins, naphthenes, and aromatics in the range of five-to-twelve carbon molecules (C_5–C_{12}). Straight-run naphtha is obtained directly from the atmospheric distillation unit. Light naphtha is the fraction boiling from 30 to 90 °C and contains C_5 and C_6 hydrocarbons. Heavy naphtha is the fraction boiling from 90 to 200 °C and contains C_7–C_9 hydrocarbons, which is the favored feedstock to the catalytic reformer. Naphtha can also be sourced from the processing of heavier crude fractions in visbreaker, catalytic cracker, hydrocracker, and coker in which olefinic hydrocarbons are present (Speight 2011).

1.2.2 Hydroprocessing

Hydrotreatment is the conventional means for removing sulfur from petroleum fractions. This process is important to avoid poisoning of the reformer catalyst and

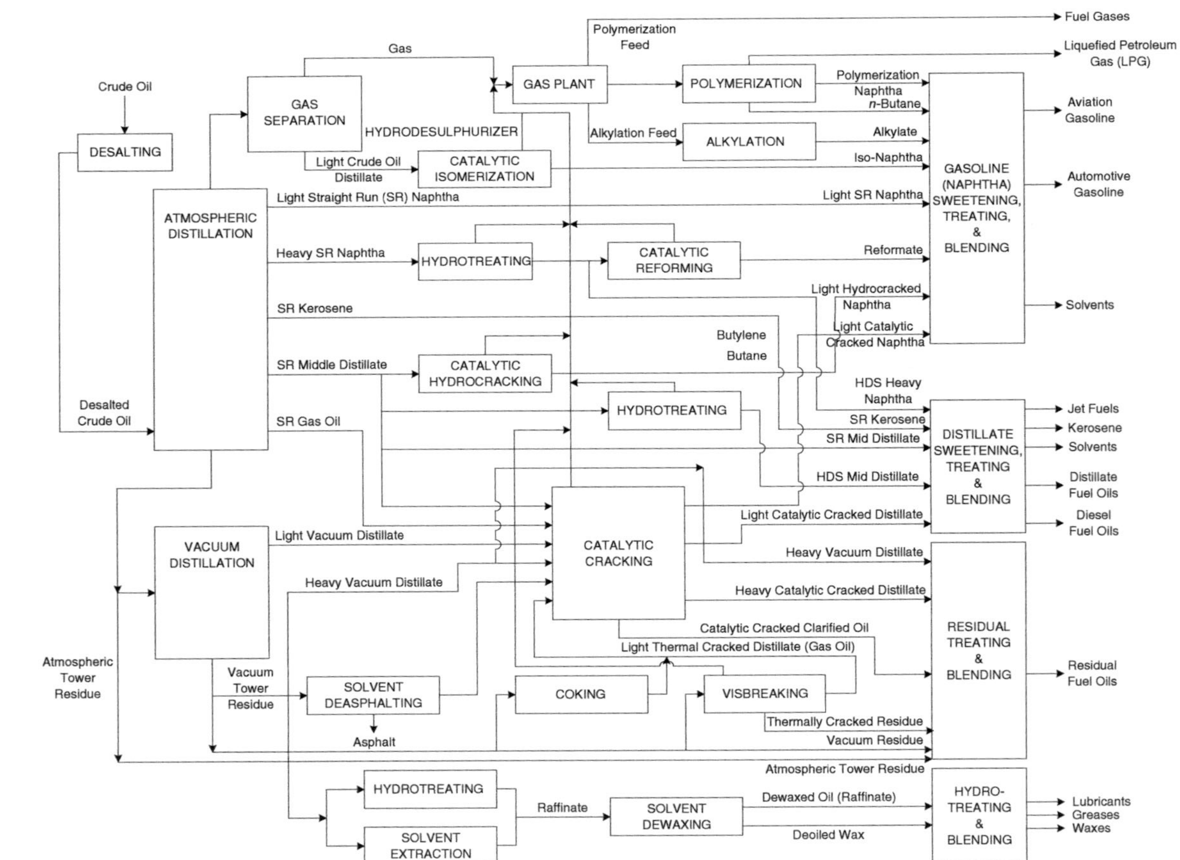

Figure 1.3 Refinery process flow. Source: Adapted from Speight (2011).

Table 1.1 Functions of major refinery processes.

Process unit	Function
Atmospheric distillation unit (ADU)	Initial separation of crude oil into the raw products of light gases, naphtha, kerosene, and diesel with the resulting residue of the atmospheric bottoms stream
Naphtha hydrotreater (HDT)	Uses hydrogen to desulfurize naphtha from atmospheric distillation, which must be hydrotreated before being sent for catalytic reforming
Catalytic reformer (REF)	Convert naphtha-boiling range molecules into higher octane reformate (i.e. reformer product) that has higher content of aromatics and cyclic hydrocarbons; an important byproduct is hydrogen released during reaction which is used either for hydrotreating or hydrocracking
Fluid catalytic cracker (FCC)	Upgrades heavy petroleum fractions into more valuable lighter products
Hydrocracker (HCR)	Uses hydrogen to upgrade heavier fractions into more valuable lighter products
Visbreaker (VIS)	Upgrades heavy residual crude oils by thermal cracking into more valuable lighter products with reduced viscosity
Coker (COK)	Converts very heavy residual crude oils into gasoline and diesel fuel with petroleum coke as residual product
Isomerizer (ISO)	Converts linear petroleum molecules to higher-octane branched molecules for gasoline blending or as alkylation feed
Alkylation unit (ALKY)	Reacts low molecular-weight olefins with an isoparaffin to form higher molecular-weight isoparaffins

to meet environmental legislations on combustion gas emissions. The feedstock is passed together with hydrogen-rich gas (usually above 75% of hydrogen by mass), over a fixed bed of catalyst under conditions that depend mainly on the feedstock properties and desired product specifications. Hydrodesulfurization consumes hydrogen and generates hydrogen sulfide according to the general reaction:

$$\text{R–S–R} + 2\text{H}_2 \rightarrow 2\text{R–H} + \text{H}_2\text{S}$$

where R represents an alkyl group and S represents a sulfur atom. The severity of a hydrotreater depends on the amount and types of sulfur compounds in the naphtha feed, which in turn are determined by the crude oil source. Characterization of sulfur compounds in naphtha is particularly difficult due to extremely low concentrations. The sulfur composition in a blend (60% straight run and 40% hydrocracked naphtha) is almost the same as straight-run naphtha since sulfur contribution from hydrocracked naphtha is negligible (Ali 2004a).

Catalytic naphtha hydrotreatment can simultaneously accomplish desulfurization, denitrogenation, and olefin saturation. Lower boiling compounds are desulfurized more easily than high boiling ones. Reactivity decreases with increasing

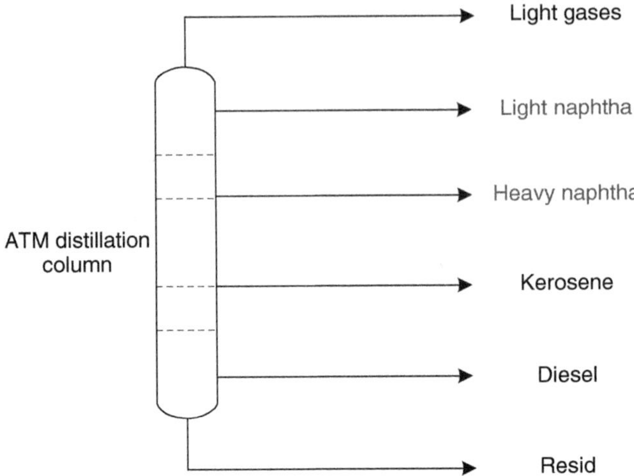

Figure 1.4 Fractions from atmospheric crude distillation unit.

Table 1.2 Naphtha hydrotreater: typical yields.

Component		Yield (weight)
Feed	Naphtha	1.0000
	H_2	0.0080
	Total	1.0080
Products	Acid gas	0.0012
	H_2-rich gas	0.0110
	LPG-rich gas	0.0058
	Desulfurized naphtha	0.9900
	Total	1.0080

molecular size. Products from the naphtha hydrotreater are generally acid gas, hydrogen-rich gas, liquefied petroleum gas- or LPG-rich gas, and desulfurized naphtha. Table 1.2 presents the typical yields in terms of weight fraction for feed and products of a naphtha hydrotreater (Parkash 2003c). The desulfurized naphtha from the hydrotreater can also be categorized as light and heavy.

The hydrodesulfurization of organosulfur compounds is exothermic. The amount of heat released increases with the number of moles of hydrogen consumed. This heat of reaction can increase the reactor temperature by 10–80 °C at nominal operating conditions depending on the feedstock (Ali 2004b). The conditions typically used to hydrotreat straight-run feedstock are mild, whereas treating cracked feeds (or blends of cracked and straight-run feeds) requires more severe conditions. The main operating variables are temperature, hydrogen partial pressure, and space velocity. In general, an increase in temperature and hydrogen partial pressure increases the

Table 1.3 Sulfur recovery unit yields.

Component		Yield (weight)
Feed	H_2S gas	1.0000
	Total	1.0000
Products	Sulfur	0.8478
	Loss	0.1522
	Total	1.0000

reaction rates of sulfur and nitrogen removal, while an increase in space velocity has the reverse effect.

1.2.3 Sulfur Recovery

The hydrogen sulfide generated in the hydrotreater is sent to sulfur recovery unit before it is burnt as refinery gas. The conversion of hydrogen sulfide to elemental sulfur is necessary to minimize atmospheric pollution by sulfur dioxide. This is in line with environmental regulations, which mandate the recovery of 99% or more of the sulfur in refinery gas (Gary and Handwerk 2001). Sulfur recovery unit operates based on the Claus process, which proceeds as follows:

$$\text{Burner: } 2H_2S + 3O_2 \rightarrow 2H_2O + 2SO_2$$

$$\text{Reactor: } 2H_2S + SO_2\ 2H_2O + 3S$$

One-third of the H_2S is converted to SO_2 by combustion, which is then combined with the remaining two-thirds and passed over a catalyst where molten sulfur forms and is separated from the gas stream. This sulfur is sold to generate additional revenue. The gas stream is cooled by steam generation and passed over another catalyst bed. This cycle is repeated for as many as four catalyst beds in some instances. The gas stream leaving the sulfur recovery unit still contains H_2S and/or SO_2, which requires further treatment. Table 1.3 shows the product yields from a sulfur recovery unit (Parkash 2003d).

1.2.4 Reforming

The continuous demand of today's automobiles for high-octane gasoline has stimulated the use of catalytic reforming to produce high-octane reformate from desulfurized naphtha without changing the boiling point range, as well as to provide hydrogen required for hydrotreating. The typical feedstocks to reformers are heavy straight-run naphtha and heavy hydrocracker naphtha. These are composed of four major hydrocarbon groups: paraffins, olefins, naphthenes, and aromatics (also referred to as PONA). The main function of a reformer is to convert paraffins and naphthenes into aromatics, subsequently producing high-octane reformate. Typical

Table 1.4 Catalytic reforming: feedstocks and products.

RON Component	Feed (vol%)	Product (vol%)
Paraffins	30–70	30–50
Olefins	0–2	0–2
Naphthenes	20–60	0–3
Aromatics	7–20	45–60

reformer feedstocks and products have the PONA analyses shown in Table 1.4 (Gary and Handwerk 2001).

Paraffins and naphthenes undergo two types of reactions in being converted to higher octane components: cyclization and isomerization. The ease and probability of either of these reactions occurring increases with the number of carbon atoms in the molecules. It is for this reason that only heavy straight-run naphtha is used for reformer feed. Light straight-run naphtha is largely composed of lower molecular weight paraffins that tend to crack to butane and lighter fractions, thus uneconomical to process in a catalytic reformer. Hydrocarbons boiling above 204 °C are easily hydrocracked and cause an excessive carbon laydown on the reforming catalyst.

The desirable reactions in a reformer that lead to forming aromatics and iso-paraffins mainly involve the following: (1) isomerization of *n*-paraffins to iso-paraffins, (2) dehydrocyclization of paraffins to aromatics, (3) dehydrogenation of naphthenes to aromatics, (4) saturation of olefins to form paraffins which then react as in isomerization in reaction (1.1) and dehydrocyclization in reaction (1.2); aromatics are essentially left unchanged (Maples 2000). Undesirable reactions are dealkylation of side chains of naphthenes and aromatics besides the cracking of paraffins and naphthenes. Table 1.5 shows typical reformer yields for three values of research octane number (RON) (Parkash 2003a).

1.2.5 Isomerization

The octane numbers of light straight-run naphtha can be improved by isomerization to convert normal paraffins of C_5 and C_6 to their isomers. This operation results in a significant octane increase because *n*-pentane has a RON of 61.7, whereas the RON of iso-pentane is 92.3 (Gary and Handwerk 1994).

Equilibrium conversion to isomers is enhanced at lower temperatures, hence a reactor temperature of 98 to 205 °C is desirable. At these low temperatures, a very active catalyst is necessary to provide a reasonable reaction rate. Catalysts used for isomerization contain platinum on various bases. Small amounts of organic chlorides are injected continuously to maintain high catalyst activities. This leads to the formation of hydrogen chloride in the reactor, which necessitates the feed to be free of water and other oxygen sources so that catalyst deactivation and potential corrosion problems can be avoided. An atmosphere of hydrogen is used to minimize carbon deposits on the catalyst but hydrogen consumption is negligible (Gary and

Table 1.5 Catalytic reforming: typical product yields in weight fraction.

RON component	96	100	102
H_2	0.0193	0.0310	0.0320
C_1	0.0085	0.0120	0.0140
C_2	0.0138	0.0200	0.0230
C	0.0269	0.0290	0.0330
iC_4	0.018	0.0170	0.0190
nC_4	0.0228	0.0230	0.0260
iC_5	0.0276	—	—
nC_5	0.0184	—	—
C_5+	—	0.8680	0.8530
C_6+	0.8447	—	—
Total	1.0000	1.0000	1.0000

Table 1.6 Isomerization yields.

Component		Yield (weight fraction)
Feed	Light naphtha feed	1.0000
	Hydrogen	0.0040
	Total	1.0040
Products	Isomerate	0.9940
	Gases	0.0100
	Total	1.0040

Source: Data from Parkash (2003b).

Handwerk 1994). Slight hydrocracking occurs during isomerization, resulting in loss of gasoline and production of light gases. Light straight-run naphtha is also sold as petrochemical feedstock besides being sent to the isomerization unit. Table 1.6 shows typical isomerization yields.

1.2.6 Blending

The final stage of the refining process is blending. This is a crucial step where the various hydrocarbon components manufactured in the refinery are mixed together to make the final products sold by the refinery. The final blend recipes depend on the quality of the available components and on customer requirements or specifications. All blended products are tested before they are sold to ensure that they meet the specifications.

1.3 Overview of Refinery Optimization Modeling

The refinery design process can be described as consisting of three stages, namely, synthesis, analysis, and optimization.

- Synthesis involves identifying the possible need for a product, e.g. by conducting market feasibility study.

1.3.1 Refinery Optimization Systems, Techniques, and Tools

A wide range of model-based systems and tools, typically computerized with associated modeling strategies, are now pervasively available and increasingly used to help design refineries in meeting desired requirements and intended applications. As depicted in Figure 1.5, the major model-based related (or assisted) elements are briefly described as follows (Engell 2007):

- regulatory control supported or enhanced with advanced process control (APC) techniques such as the popular model predictive control (MPC) systems;
- real-time optimization (RTO) that can be implemented for both static (i.e. steady state with certain criteria for such a certification, which is the typical or traditional practice) and dynamic conditions;
- instrumentation such as sensors, analyzers, and actuators which deliver digital information at an enhanced rate;
- programmable logic controllers (PLC), distributed control systems (DCS), personal computers, and real-time servers, which are now considerably inexpensive, thus more available to buy (even in excess of required capacities) and maintain;
- new online analysis technology (including analyzers) such as near-infrared spectroscopy (commonly known by its abbreviation of NIR);
- software that supports devices, systems, and technologies enabled by (Industrial) Internet-of-Things or more popularly abbreviated as (I)IOT, which capitalizes on the capability afforded by the Internet including high-performance tools for storing and accessing data (e.g. in the cloud or at the edge);
- new tools as means of communication that are increasingly rapid and reliable from transmitting initial input data up to integrating with management level (e.g. at the headquarters);
- software development efforts that are continually reducing the need for proprietary modeling systems requiring the expertise, knowledge, or skills of a specialist;
- new powerful computational concepts, methods, and techniques for model development and solution such as machine learning (e.g. neural network), which has gained a lot of attention recently (besides mathematical programming, constraint programming, fuzzy logic, or others).

A number of commercial optimization tools are routinely used in the industry that typically include the following:

- PIMS (Process Industry Modeling System by Aspen Technology): Technique called sequential linear programming (SLP) augmented with recursion methods

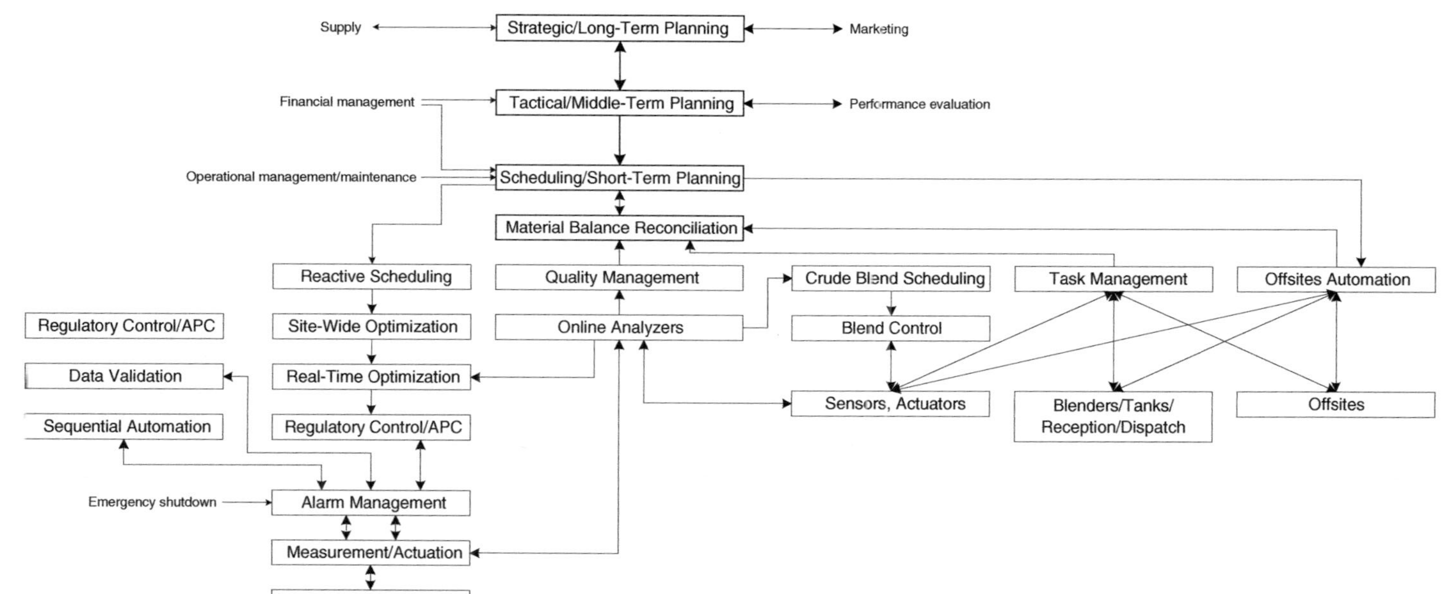

Figure 1.5 Model-based related elements for typical refinery decision-making support systems.

is employed to enable handling some nonlinear functions (e.g. bilinear terms that arise due to mixing process operations) through an iterative procedure; spreadsheet-based user interface; and reporting;

- Excel Solver (add-in by FrontLine used in Microsoft Excel): Uses spreadsheet format for inputs/outputs;
- Sequential quadratic programming (SQP) as employed (also pioneered) through the deployment of dynamic matrix optimization (DMO) technique: Uses algorithm based on derivatives of functions to search for optimality; equations are written in standard Fortran format with extensive linking to execute optimization functionalities.

1.3.2 Modeling for Advanced Process Control

APC is an area that has long experienced a widespread practice of implementing model-driven or model-assisted technology (Lee et al. 2018). APC applications in refineries are mainly concerned with real-time multivariable control as based on (largely) empirical models of process units. Such models can be incorporated with dynamic RTO to improve the setpoints computed for implementation in APC, most commonly through applying MPC systems (Kadam and Marquardt 2007; Pontes et al. 2015).

The availability of sufficient degrees of freedom for modeling, control, and optimization supported by excellent instrumentation for measurements has allowed refineries to benefit significantly from these model-based techniques and tools. The advantages can be exemplified (in whole or part) through easier, faster, safer, and even greener (i.e. more environmental friendly) process unit management afforded by more effective response. The resultant stability of operating conditions also improves handling of changes in the production environment such as variations in feedstock quality and weather conditions.

Consequently, these granted refinery operations to maximize throughput of higher feed flow rate resulting from being able to push multiple process constraints in an optimal manner to higher levels. At the same time, increased process capability due to reduced process variabilities permitted higher stream yields in meeting market demands particularly for high-value products. As a result, refineries improve energy and general operation efficiency through less consumption of utilities and chemicals as well as catalysts, lower losses due to manufacturing-related issues; and decreased fluctuations in quality and the associated giveaways in product properties.

Modeling using APC contributes significantly to a large part of the potential benefits derived in which the benefits can be quantified from refinery revamp projects. Thus it is prioritized to carry out such value-improvement projects. Some of the popular commercial APC software systems widely used in refineries worldwide include Aspen Technology's DMC (which historically stands for Dynamic Matrix Control) and Honeywell's Robust Multivariable Predictive Control Technology (RMPCT) (Morari and Lee 1999). Trade information up-to-date on currently operating APC systems is available in references such as Hydrocarbon Processing, a monthly periodical which also publishes an annual supplement (called Advanced Process Control and Information Systems Handbook).

APC has been a mainstay in the refinery suite of modeling tools and applications for the past two decades or so (Forbes et al. 2015). In comparison, optimization modeling tools for online (real time) or offline purposes have not been used as commonly, extensively, or pervasively relative to APC although arguably the trend has gained more widespread acceptance in more recent times (Lee 2011).

Despite its many advantages, a possible objection to APC is that even erroneous setpoints are simply implemented quickly when the most difficult part of refinery operations is to ensure that setpoint values as accurate as possible are established in the automated controls. Thus, RTO, also known as online optimization systems, is employed for determining the optimal values for the setpoints sent to the APC by executing online calculations (as briefly reviewed next).

1.3.3 Modeling for Real-Time Optimization

Compared to APC, there are relatively fewer commercial RTO applications; even less common is a plantwide operational optimizer (with the level of detail as that of a typical RTO), which is currently used in the downstream petroleum processing industry. Conventional RTO models are required to generate solutions within minutes of execution (after achieving steady state as determined by certain appropriate criteria). At the same time, the models are updated with economic and technical conditions data of process units in a continuous and automated way. The latter necessitates reliable and reasonably fast digital systems capable of collecting and validating the data including to verify steady-state conditions in developing the processing unit flowsheeting models.

In a similar way, a number of process simulation packages incorporate RTO-like features of general optimization capabilities in addition to offering combined or hybrid first-principle and empirical modeling options. Since the models used for the optimization must be able to integrate all of the system's constraints and use the degrees of freedom available, the resulting optimization models formulated can be complex. The complexity is manifested particularly by the requirement to handle over hundreds of thousands of equations in which some are possibly nonconvex (e.g. bi-/tri-linear or exponential) as well as millions of variables, all subject to a simultaneous solution strategy to be devised and executed in arriving at some meaningful (if not practical) results.

It is noteworthy that a standard rigorous approach of RTO-based modeling typically involves the following sequential procedure (as summarized in the flowchart shown in Figure 1.6). The first step involves periodically executing an algorithm, which is generally based on nonlinear optimization that gathers data from steady-state operations. The second step performs data reconciliation between the actual plant and model values. The third step consists of adapting or updating some model parameters so that the model results match the reconciled data to an acceptable degree of accuracy. The fourth step computes optimum target values as improved setpoint candidates for identified (i.e. preselected or preconfigured)

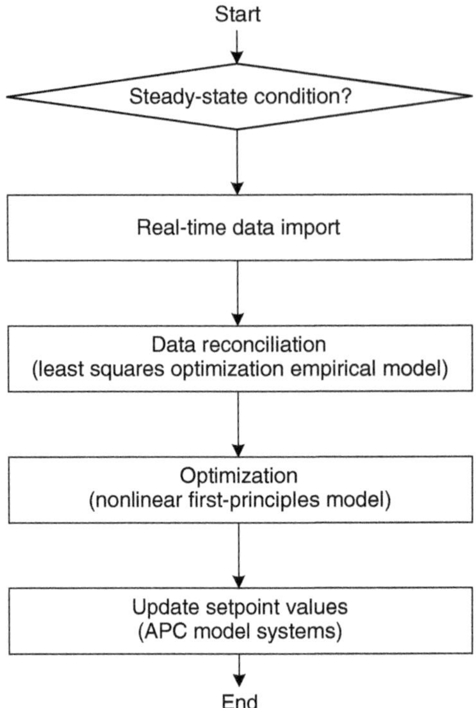

Figure 1.6 Standard cycle procedure for typical real-time optimization modeling systems (online and closed loop).

process variables as handles to optimize (mainly for local optima) the process units or the entire refinery in general. The final step entails modifying by implementing the necessary process setpoints. The cycle is then repeated when a steady-state criterion is verified to be attained (Bodington 1995; Moro 2003).

Recent developments in the petroleum refining world include efforts to implement simultaneous optimization of models for multiple units in real-time mode. An ultimate goal is to enable online economic optimization (i.e. with commercial significance or impact) of the total refining process from the front-end (including berth- or dock-related activities) to the back-end (including end-products distribution). A crucial enabler for successful applications to problems of industrial scale and significance is the availability of powerful nonlinear optimization algorithms (de Prada et al. 2017).

Conventional RTO software packages are equation-oriented systems that make use of robust large-scale nonlinear programming algorithms as the solution engine (e.g. SQP) such as the implementation in the commercial package called ROMeo (Rigorous Online Modeling with equation-based optimization) (AVEVA 2021). A user can expect to find in an RTO package such standard features like combination of the underlying mathematical formulation and solution engine with an interface (typically GUI or graphical user interface). The combined components of the package or systems are designed to capture and show real-time plant data and economic objective function values, which attempt to depict a replica (i.e. a model) of the plant operations in as close as possible a manner.

A modeling-aided management system such as RTO, which is coupled to APC offers a capability that allows users (i.e. plant engineers as well as the management besides the modelers especially those personnel at the site) to have a reliably rigorous, precise, and accurate model of the process units in an operating facility to optimize operations on a plant-wide basis by determining optimal process setpoints for the plant control systems. In doing so, such a model-based system also assists in determining the cause and source (or location) in which operating bottleneck problems and challenges lie. A single model with a modern user-friendly GUI approach (such as that in the manner of RTO), instead of separate models catering for multiple uses including process simulation, data reconciliation, and operational optimization, facilitates to develop, deploy, and sustain a model with reduced cost.

Notwithstanding significant continuous progress as mainly found reported in both trade periodicals and academic journals, which includes a recent partnership between AVEVA and ExxonMobil Research and Engineering Co. (EMRE) to enhance ROMeo's capability, it remains arguable that RTO still lacks general industrial acceptance (not least, relative to APC). The reasons could be due to certain practical and also theoretical limitations. Some of these drawbacks generally pertain to the levels of detail in characterizing the feed streams, developing the process unit models, and detecting the existence (or nonexistence) of true steady state (the latter applies particularly to plants with reasonable disturbance).

Nonetheless, substantial noteworthy effort has been undertaken to address these issues for both the system modeling as well as the solution parts of RTO technology. Improvements are partly reflected from industrial implementations of RTO particularly for ethylene processing plants for petrochemical production besides the petroleum refining industry (Shobrys and White 2000; Moro 2003; Karuppiah and Grossmann 2006). Nonetheless, existing commercial RTO packages have been thought of as somewhat dated (notwithstanding ongoing modernization initiatives to incorporate state-of-the-art features, which include utilizing cloud and edge computing technologies). Some examples along this line are exemplified in packages enabled by dynamic optimization-based modeling and solution techniques such as gPROMS software by Process Systems Enterprise (now owned by Siemens) or GDOT software by Aspen Technology (now owned by Emerson).

1.3.4 Modeling for Process Simulation

Commercial model-driven process simulators are now largely considered as a common omnipresent basic available tool required to enable and assist refinery process engineers mainly to perform detailed material and energy balances. Subsequently, the results (mostly available for steady-state condition but increasingly also for dynamic condition) are requisite for a variety of other tasks and purposes as part of design, operation or production, and maintenance or sustainment support activities. Some examples of the latter (i.e. for maintenance applications) are more and more used for debottlenecking studies on unit operations for plant revamp and rejuvenation exercises.

Notable examples of general-purpose simulators include Aspen Plus (Aspen Technology 2021b); UniSim (Honeywell 2021); VMGSim (Virtual Materials Group (VMG) 2021) for which different versions have been customized (called iCON for PETRONAS, the Malaysian national oil company) and now acquired by Schlumberger (called Symmetry); PRO/II (formerly of Invensys and now acquired by Schneider Electric); and Petro-SIM (formerly of KBC and now acquired by Yokogawa). A number of these packages also offer tailored or bespoke detailed models for specific refinery processes such as fluidized catalytic cracking (FCC), catalytic reforming, and hydrocracking. On the other hand, commercial specialized simulation tools are also available to perform realistic simulation-based modeling for certain process units in which detailed diagnostic studies are necessary. In this realm, some of the examples pertaining to refinery-based applications include those that need to incorporate detailed reaction kinetics in the modeling of conversion units particularly for catalytic reforming, FCC, and hydrocracking units.

In addition, process simulation have been integrated as part of the routines in certain plant optimization models. Such integration are performed in either online or offline mode—typically the former is executed on continuous real-time basis while the latter on longer-term advisory or study basis. These integrated models typically employ the latest process information extracted or imported from process simulation models to update existing process plant models (Hallale et al. 2006).

As an example, Aspen Technology, a perennial market leader in provisioning of modeling and optimization software, systems, and support services particularly for refinery planning and scheduling, has extended its models to make available detailed reactor representation with rigorous kinetics modeling for optimizing hydrogen production (Aspen Technology 2021a). Such models, which are dubbed as hybrid models since they combine both first-principles and empirical modeling techniques coupled with artificial intelligence and analytics algorithms, have enabled a better understanding of process operations under different hydrogen feed conditions. The hybrid reactor simulation models developed in AspenPlus can also be linked to and employed in Aspen PIMS with automatic updates available, thus allowing the combined models to perform enhanced detailed planning and scheduling activities as well as economic evaluation (Beck and Munoz 2020).

1.3.4.1 Modeling for Dynamic Simulation

A related development concerns modeling for dynamic process simulation. Several software packages have resulted from more than three decades of research and development (R&D) intertwined with systems design or customization. Applications of unit-wise and plant-wide dynamic simulation can be readily adopted as part of an extensive modeling suite now available to assist with optimal decision-making in refinery operations. Some examples of early equation-based modeling software systems include DIVA (Kröner et al. 1990), ABACUSS (Allgor et al. 1996), and SPEEDUP (Perkins and Sargent 1982) – the latter forms a base that has been further developed and evolved to be a state-of the-art current cutting-edge package called gPROMS (Barton and Pantelides 1994) by Process Systems Enterprise (a spinoff company of Imperial College London's Department of Chemical Engineering, which since has been acquired by Siemens) (Siemens 2021).

Such large-scale dynamic system simulation software allows a modeler or an engineer to focus solely on formulating and thereafter implementing the model developed. An advantage of doing so is that it generally removes or obviates the modeler/engineer from concerns about solution algorithms (such as infeasibility due to numerical issues caused by parameter or variable scaling) and code-related issues (such as syntax for generation, compilation or interpretation, and debugging). This modeling aid greatly permits the modeler/engineer to increase productivity while assisting to ensure the dynamic simulation feasibility. Relevant activities regarding such refinery use include designing, verifying, and analyzing parametric sensitivity for control and safety interlock systems as well as investigating operational events of start-up, changeover, and shutdown.

1.3.4.2 Modeling for Operator Training Simulation

Refinery training simulators based on dynamic simulation modeling technique (as discussed in Section 1.3.4.1) are now an essential tool to train new operators especially panel operators manning the control rooms besides to maintain and update the operators' knowledge and skill regularly in reacting to exceptional circumstances. Although (panel) operators now are exposed to operational incidents much less frequently than in the past (mainly due to control room reorganization), the increased scope, level, and reliability of automation (which is also partly exacerbated by the Covid-19 virus outbreak pandemic) have led to requirements for increased duration and alternative suitable methods for operator training. Thus to guarantee a refinery's safety in the situation of automation systems failure, which can lead to plant failure altogether, it is imperative to enhance operators' readiness through enhanced training level including by adopting a model-based approach.

We can represent realistic operating conditions using dynamic simulation models in a manner akin to how flight simulators are used for aircrew training (e.g. for civilian aircraft or military air force). The similarity of the roles is particularly apparent to allow modeling of transient time period in the event that follows after an incident or operational change. We can also expect state-of-the-art dynamic simulators to be able to model several possible scenarios including to evaluate historical data of past operator's real-time reactions.

Models for training simulators typically include operator consoles (which are identical to those one may expect to see in the control rooms) as a feature linked to the DCS. Modeling in this way enables access to initial states of the process variables. Indeed, training simulators and their comprehensive use are a showcase of modeling capability in supporting high competence of safety assurance in refinery operations.

1.3.5 Modeling for Planning and Scheduling

Modeling of refinery planning and scheduling serves to represent, predict, and prescribe the establishment of a detailed manufacturing activities plan for a plant across various time scales (ranging from seconds or minutes up to several years). Planning and scheduling enable and support enterprise supply and marketing functions of a company by defining specific activity timings in producing the requisite product volumes that fulfill the company's objectives. Based on requirements stipulated by the company operations, the main goal of refinery planning and scheduling is in meeting the demands of primary outlets of national and international retail sales (Pompéi 2001). A side goal of the production plans and schedules also serves to minimize any associated byproducts generated since typically they are of uneconomical values.

Thus, modeling for refinery planning and scheduling systems is vital particularly to improve feeds as throughput to the process units in realizing potential large returns or margins and saving costs significantly. But a main complicating challenge in modeling refinery operations lies in representing the operation modes of the process units involved, i.e. that can be not only mostly continuous in nature but also batch or semi-batch (or semi-continuous). Such operating considerations are on top of a need to deal with the existing complexity involved in handling the various crude oils feed mix and product grades. In addressing these issues, the modeling effort entails adopting both the more conventional continuous and discrete types of decision variables and the corresponding constraints as adequately espoused elsewhere (Karuppiah et al. 2008). Doing so allows modeling with greater flexibility besides incorporating practical problem representation in the model formulation as will be illustrated throughout the book.

As shown in Figure 1.7, planning and scheduling activities in a refinery are made up of several sections. These sections can be represented conveniently as separate modules of optimization models, namely, crude oil movement and blend scheduling (at the refinery front-end), production planning and scheduling (refinery middle-end), and product blending and distribution scheduling (refinery back-end) (Jia and Ierapetritou 2003; Reddy et al. 2004; Méndez et al. 2006).

The front-end plans and schedules involve blend scheduling optimization of crude oil mixtures (i.e. crudes for short) that include handling arrival events of crudes such as unloading from vessels, transfer to storage tanks, and blending and charging of crudes as feedstock for processing in crude distillation units. The middle-end plans and schedules address process unit operations of the intermediate streams including their inventory control from one unit to another. The process operations involve reactions to modify and attain the desired output such as product properties (e.g. higher octane number) as well as separation to increase product purity (e.g. lower contaminant concentrations), which can entail a need for handling effluents. The back-end plans and schedules deal with product blending and dispatch or distribution in which the latter largely involves logistics optimization around end-products lifting for shipping delivery to customers (i.e. demand centers or points) and including inventory control that serves to prevent stock out.

Refinery commercial success depends to a great extent on how the associated plans and schedules are modeled for development, deployment, and sustainment. Thus, it is imperative for a refinery to be able to put in place a robust and reliable as well as automated process of modeling and computing in generating optimal or feasible plans and schedules. Typically, a database, e.g. one based on SQL

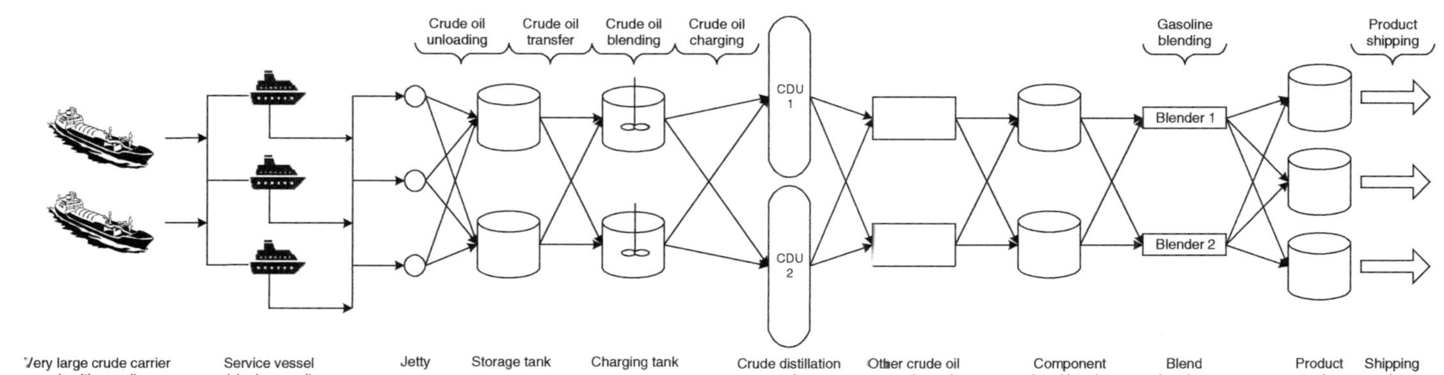

Figure 1.7 Schematic for optimization modeling of refinery end-to-end encompassing crude oil blend scheduling (at front-end), process planning and scheduling (middle-end), and product blending and delivery (shipping) scheduling.

(structured query knowledge) is used particularly within a short-term scheduling model to ensure that a refinery implements the generated plans and schedules in a consistent manner between one to the other with the latter (i.e. results from the scheduling activity) involving more details or granularity than the former (i.e. results from the planning activity).

The consistency in data and its validity are ensured particularly on yields, costs, and throughputs. Such undertaking necessitates significant requisite effort in maintaining, establishing, and integrating the database with the refinery-wide information systems encompassing both IT (information technology) and OT (operating technologies) including management-centric dashboards.

In view of the drive for increased or enhanced digitalization spurred by the Industrial Revolution 4.0 (IR4.0) paradigm and the onset of the Covid-19 (coronavirus) pandemic, model-based tools for decision-making support systems have seen a surge of interests particularly but not limited to the activities of optimizing refinery plans and schedules. The actual situation is not something entirely new because even previously, refineries face amplified pressures constantly due to intensified operating challenges brought about by tightened market, regulatory, and environmental requirements (also sometimes referred to as the so-called "triple bottom-line") as well as heightened priorities for social and governance aspects.

Ongoing recent technical advances have generated renewed interests in computer-aided model-based decision-making tools to improve refinery planning and scheduling activities continually. In particular, the following refinery optimization developments have enabled more benefits to be reaped with increased speed and scale:

- greater operational complexity in raw materials handling due to pressures of attaining economic advantage of processing cheaper challenged crudes (with typically more diverse crude oil mixtures);
- quicker and more rapid supply chain responsiveness to market dynamics to capture more value;
- reduced operating degrees of freedom arising from a greater number of constraint, e.g. more stringent product specifications, stricter environmental requirements, and tighter tankage limitations;
- generally easier, faster, and more accurate (relatively) problem formulation through adopting increasingly powerful state-of-the-art modeling techniques (e.g. logic-based constraint programming, artificial-intelligence-based expert systems, and hybrids of mixed-integer optimization involving combined mixed-integer programming with disjunctive programming);
- by a similar token to the previous point (i.e. in the same way as afforded by the availability of more sophisticated models), more efficient solution and implementation or deployment of model-based optimization for decision-making support system;
- more advanced computing and computational capacity afforded by the increased availability of more powerful hardware and more efficient algorithms;
- massive refinery data availability (i.e. in terms of volume, velocity, variety, veracity, and value which collectively describes the "big data" phenomenon) as well as that

related to the parent company or enterprise networks, which has given access to requisite information instantly and pervasively through internet and web-based technologies or infrastructures (e.g. cloud computing or Industrial IoT components in general).

The situation has enabled complex planning and scheduling problems with solution methods and strategies that can cater for discrete decisions. Such models involve both continuous and integer (often binary) types of decision variables. Thus, the application necessitates handling an exponential number of feasible plans and schedules with their associated permutations. In particular, the mathematical formulation and model development undertaking give rise to combinatorial optimization of mixed-integer programming with tools and techniques that can now afford such problems to be solved within tractable and practical computational effort and execution or run time.

The techniques of advanced model-based optimization coupled with powerful and scalable yet flexible object-oriented programming and graphical interfaces have been largely instrumental in enabling the practical use of these resource-and-data-intensive systems. There is demonstrated great potential for enhanced practical use of planning and scheduling systems that can benefit from the joint employment of machine learning-based analytics (such as artificial neural networks or data mining with decision trees) and artificial intelligence-based cognitive techniques (such as expert systems) coupled with the latest clever mathematical algorithms made possible by high-performance computer hardware (as also mentioned elsewhere). In view of current developments, model-based computer systems are increasingly if not pervasively substituting the function of many manual (nonautomated) procedures, which typically rely on spreadsheet-based tools presently used for planning and scheduling activities of complex refineries (particularly in handling multiple crude oils processing). These activities frequently entail making decisions for multiperiod problems that involve optimization to sequence crude oil runs with consideration for product blending rules, allocate storage and charging tanks, determine operating conditions of the process units and their associated auxiliary or supporting units, devise product blending formulations, and sequence the movements of finished products for distribution to depots or ports as well as to be ready for shipping to sales points of end customers.

1.3.5.1 Systems Implementation

Planning and scheduling systems are at the heart of refinery IT systems as well as increasingly its OT systems. Thus it is imperative that data for the plans and schedules are continuously kept up to date. The main data required comprises tank levels (as a function of time) especially those related to inventory, operating conditions of process units, component and product qualities, and major equipment availability besides business-related data such as marketing requirements (or situations) as well as forecast of external supply or deliveries and product lifting requirements.

Refineries require robust, reliable, and secure communications network that include auxiliary systems such as an associated database to store plant-wide data for the necessary or requisite processing and manipulation steps of extraction, transformation, and loading to the relevant applications. This infrastructure is deemed as crucial elements or components yet with potential room for major improvements particularly in view of harnessing the pervasive "big data" available today in deriving greater value from better decision-making activities. Planning and scheduling functions are now adopted as decision support systems. However the scope of such systems typically do not guarantee whole refinery optimization of the operations, mainly due to the challenge of high complexity and consequently a long model solution time (i.e. runtime).

It is typically practiced in the industry to develop separate modeling applications on crude oil charging and blend scheduling (at refinery front-end) and that on planning and scheduling of product blending and delivery (at refinery back-end). Plans and schedules tend to be obsolete (i.e. no longer relevant to business situations) fairly rapidly due to uncertain product demands and other market-related factors or even sentiments. For the latter problem, the uncertainty arising from fluctuating product demands and other changing parameters of the market often render the plans and especially the schedules to be obsolete very quickly. Thus, such problems (particularly the schedules) are mainly used for optimization studies on the products tankage.

1.3.5.2 Optimization of Crude Oil Scheduling

As crude oil purchasing costs account for more than two-thirds (around 80%) of a refinery's turnover, we can achieve significant savings by optimally performing the scheduling operation of charging crude oil blends to the primary (atmospheric) distillation units (Kelly and Mann 2003). With premium crude oils experiencing reduced supplies and increased prices or costs, a significant operational optimization issue that afflicts refineries today revolves around how to feasibly (if not optimally) exploit greater margins from employing low-cost crude oils or the so-called "challenged crudes", i.e. to increase profits by using varying blends of premium and challenged crude oils. However, various operational issues are expected to arise related to processing these challenged crudes that have high contents of contaminants including aromatics, sulfur, and various other high residues. Therefore, refiners need to strategically identify an optimal mix of both premium and challenged crude oils for processing that does not compromise profit margins and with as few operational challenges as possible to this crude oil blend scheduling optimization problem.

It is not surprising that the functions, duties, and responsibilities of refinery crude oil schedulers (i.e. optimization application users) are not only ever more demanding and pressing now but increasingly complex and convoluted, which entails increasingly higher risks at stake operationally, financially, and politically. It is requisite that the users monitor and match crude oil movement to market demand as well as that of the refinery overall operation.

Under normal circumstances of handling crude oil receipts or arrivals, the users assign destination tanks for each of the crude oil shipment parcels. This task also includes blending and charging the received crude oils to the distillation units (i.e. CDU) as necessary in meeting mainly yield and quality targets or various other specifications on the fractionation products. However, due to pressures arising from timing or deadlines and unfavorable inventory availability or flexibility, the users tend to rely on their experience (if not intuition) that prompts them to select the first feasible solution computed by existing non-rigorous model-based tools (e.g. spreadsheets using Excel typically). As a result, there is significant loss of economic and operational opportunities to improve such planning and scheduling decision-making activities.

Automated, reliable, and robust model-based refinery crude oil blend scheduling optimization system is valuable and needed for the foregoing reasons with potential of enhancing crude oil blend options by considering various possible (and permissible) charge mixtures (i.e. combinations or permutations of crude oil sources or types in optimum proportions) particularly in capturing value from processing high-quality crude oils which are mixed with less expensive feedstock in an economically optimum manner. Other associated benefits or returns that such advanced optimization systems can offer include increased throughput (i.e. processing output) and raw material resource utilization, reduced yield and quality giveaways, enhanced control and predictability of downstream production, and decreased costs (e.g. inventory holding, storage, and demurrage).

1.3.5.3 Refinery Management

A real challenge in using modeling tools lies in ensuring that we can get realistic results compared to an existing possibly manual approach. With the present drive for harnessing the power and resultant benefit of increased digitalization, it is imperative to be able to showcase and continually improve the value of modeling (or a model-driven approach, in general) to refiners. The outcome of improvement brought about through such efforts typically requires minimal additional bespoke or tailored custom-built features with a potential incentive that they can be directly adopted to meet actual refinery operations while the derived benefits can be straightforwardly quantified (e.g. in monetary sense). Some of the advancements include linking model-based optimization tools with real-time databases for which such automation permits access to data and insights that could assist with planning and scheduling activities. An example entails validating outputs from (long-term) plans and (short-term) schedules enhanced with simulation results with an aim of gaining more granularity toward attaining more realistic production targets setting.

1.4 Concluding Remarks

This chapter provides an overview of the subject addressed in this book, which concerns a mixed-integer superstructure optimization approach to determine the topology or configuration for a petroleum refinery. A brief treatment of refining processes is given as background on the numerous available commercial process units considered in the approach. The chapter then proceeds to present a representative selection of staple refinery optimization modeling tools as partly exemplified by the current practice of RTO and APC. A brief exposition is also offered as regards to modeling-aided suite of techniques and approaches for efficient and effective plant management as appropriate in this age of digitalization as spurred by Industrial Revolution 4.0, big data analytics, and to a certain extent, the global coronavirus (Covid-19) outbreak, chiefly to lend a flavor to the book's subject but without attempting to be comprehensive (with references cited for a more complete exposition).

References

Ali, S.A. (2004a). Naphtha hydrotreatment. In: *Catalytic Naphtha Reforming* (ed. G.J. Antos and A.M. Aitani), 114. New York: Marcel Dekker.

Ali, S.A. (2004b). Naphtha hydrotreatment. In: *Catalytic Naphtha Reforming* (ed. G.J. Antos and A.M. Aitani), 119. New York: Marcel Dekker.

Allgor, R.J., Berrera, M.D., Barton, P.I., and Evans, L.B. (1996). Optimal batch process development. *Computers & Chemical Engineering* 20 (6/7): 885–896.

Aspen Technology (2021a). Aspen Hybrid Models. Aspen Technology. https://www.aspentech.com/en/solutions/aspen-hybrid-models (accessed 11 November 2021).

Aspen Technology (2021b). AspenPlus. http://www.aspentech.com (accessed 11 November 2021).

AVEVA. (2021). AVEVA process optimization. AVEVA Group. https://www.aveva.com/en/products/process-optimization (accessed 11 November 2021).

Barton, P.I. and Pantelides, C.C. (1994). Modeling of combined discrete/continuous processes. *AIChE Journal* 40: 966–979.

Beck, R, and Munoz, G. (2020). Hybrid modeling: AI and domain expertise combine to optimize assets. https://www.aspentech.com/en/resources/white-papers/hybrid-modeling-ai-and-domain-expertise-combine-to-optimize-assets-cxo (accessed 11 November 2021).

Biegler, L.T., Grossmann, I.E., and Westerberg, A.W. (1997). *Systematic Methods of Chemical Process Design*. New Jersey: Prentice Hall.

Bodington, C.E. (ed.) (1995). *Planning, Scheduling, and Control Integration in the Process Industries*. New York: McGraw-Hill.

Douglas, J.M. (1985). A hierarchical decision procedure for process synthesis. *AIChE Journal* 31 (3): 353–362.

Douglas, J.M. (1988). Conceptual design of chemical process.

Engell, S. (2007). Feedback control for optimal process operation. *Journal of Process Control* 17 (3): 203–219.

Forbes, M.G., Patwardhan, R.S., Hamadah, H., and Gopaluni, R.B. (2015). Model predictive control in industry: challenges and opportunities. *IFAC-PapersOnLine* 48 (8): 531–538.

Gary, J.H. and Handwerk, G.E. (1994). *Petroleum Refining: Technology and Economics*, 3e, 204. New York: Marcel Dekker.

Gary, J.H. and Handwerk, G.E. (2001). *Petroleum Refining: Technology and Economics*, 4e, 273. New York: Marcel Dekker.

Grossmann, I.E. (1996). Mixed-integer optimization techniques for algorithmic process synthesis. In: *Advances in Chemical Engineering* (ed. J. Wei, M.M. Denn, G. Stephanopoulos, and J.H. Seinfeld), 171–246. Elsevier.

Grossmann, I.E. and Trespalacios, F. (2013). Systematic modeling of discrete-continuous optimization models through generalized disjunctive programming. *AIChE Journal* 59 (9): 3276–3295.

Hallale, N., Moore, I., and Vauk, D. (2006). Hydrogen: under new management. In: *Practical Advances in Petroleum Processing* (ed. C.S. Hsu and P.R. Robinson), 371–392. New York: Springer Science+Business Media, Inc.

Honeywell (2021). UniSim – software for process design and simulation. https://www.honeywellprocess.com/en-US/explore/products/advanced-applications/unisim/Pages/default.aspx (accessed 11 November 2021).

Jia, Z. and Ierapetritou, M. (2003). Mixed-integer linear programming model for gasoline blending and distribution scheduling. *Industrial & Engineering Chemistry Research* 42 (4): 825–835.

Kadam, J.V. and Marquardt, W. (2007). Integration of economical optimization and control for intentionally transient process operation. *Lecture Notes in Control and Information Sciences* 358: 419–434.

Karuppiah, R. and Grossmann, I.E. (2006). Global optimization for the synthesis of integrated water systems in chemical processes. *Computers & Chemical Engineering* 30 (4): 650–673.

Karuppiah, R., Furman, K.C., and Grossmann, I.E. (2008). Global optimization for scheduling refinery crude oil operations. *Computers & Chemical Engineering* 32 (11): 2745–2766.

Kelly, J.D. and Mann, J.L. (2003). Crude oil blend scheduling optimization: an application with multi-million dollar benefits. *Hydrocarbon Processing* 82 (6): 47–52.

Khor, C.S. and Varvarezos, D. (2017). Petroleum refinery optimization. *Optimization and Engineering* 18 (4): 943–989.

Khor, C.S., Yeoh, X.Q., and Shah, N. (2011). Optimal design of petroleum refinery topology using a discrete optimization approach with logical constraints. *Journal of Applied Sciences* 11 (21): 3571–3578.

Kröner, A., Holl, P., Marquardt, W., and Gilles, E.D. (1990). DIVA—an open architecture for dynamic simulation. *Computers & Chemical Engineering* 14 (11): 1289–1925.

Lee, J.H. (2011). Model predictive control: review of the three decades of development. *International Journal of Control, Automation and Systems* 9 (3): 415.

Lee, J.H., Shin, J., and Realff, M.J. (2018). Machine learning: overview of the recent progresses and implications for the process systems engineering field. *Computers & Chemical Engineering* 114: 111–121.

Maples, R.E. (2000). *Petroleum Refinery Process Economics*, 2e, 264. Oklahoma: Pennwell.

Méndez, C.A., Grossmann, I.E., Harjunkoski, I., and Kaboré, P. (2006). A simultaneous optimization approach for off-line blending and scheduling of oil-refinery operations. *Computers & Chemical Engineering* 30 (4): 614–634.

Morari, M.H. and Lee, J. (1999). Model predictive control: past, present and future. *Computers & Chemical Engineering* 23 (4): 667–682.

Moro, L.F.L. (2003). Process technology in the petroleum refining industry—current situation and future trends. *Computers & Chemical Engineering* 27 (8): 1303–1305.

Nemhauser, G.L. and Wolsey, L.A. (1988). *Integer and Combinatorial Optimization*. Wiley-Interscience.

Oil & Gas Journal (Editorial) (2019). Construction costs continue to rise, operating costs decline. *Oil & Gas Journal* 117: 27–28.

Parkash, S. (2003a). Chapter 4: Gasoline manufacturing processes. In: *Refining Processes Handbook* (ed. S. Parkash), 116. Burlington: Gulf Professional Publishing.

Parkash, S. (2003b). Gasoline manufacturing processes. In: *Refining Processes Handbook* (ed. S. Parkash), 144. Burlington: Gulf Professional Publishing.

Parkash, S. (2003c). *Refining Processes Handbook* (ed. S. Parkash). Burlington: Gulf Professional Publishing.

Parkash, S. (2003d). Sulfur recovery and pollution control processes. In: *Refining Processes Handbook*, 225. Burlington: Gulf Professional Publishing.

Perkins, J.D. and Sargent, R.W. (1982). SPEEDUP: a computer program for dynamic simulation and design of chemical processes. *AIChE Symposium Series* 78: 1–11.

Pompéi, C. (2001). Functional and organizational analysis. In: *Refinery Operation and Management* (ed. R. Baker), 455–509. Paris: Editions Technip.

Pontes, K.V., Wolf, I.J., Embiruçu, M., and Marquardt, W. (2015). Dynamic real-time optimization of industrial polymerization processes with fast dynamics. *Industrial & Engineering Chemistry Research* 54 (47): 11881–11893.

de Prada, C., Sarabia, D., Gutierrez, G. et al. (2017). Integration of RTO and MPC in the hydrogen network of a petrol refinery. *Processes* 5 (1): 3.

Raman, R. and Grossmann, I.E. (1991). Relation between MILP modelling and logical inference for chemical process synthesis. *Computers & Chemical Engineering* 15 (2): 73–84.

Raman, R. and Grossmann, I.E. (1992). Integration of logic and heuristic knowledge in MINLP optimization for process synthesis. *Computers & Chemical Engineering* 16 (3): 155–171.

Raman, R. and Grossmann, I.E. (1993a). Relation between MILP modeling and logical inference for chemical process synthesis. *Computers & Chemical Engineering* 15: 73–84.

Raman, R. and Grossmann, I.E. (1993b). Symbolic integration of logic in mixed integer linear programming techniques for process synthesis. *Computers & Chemical Engineering* 17: 909–927.

Reddy, P.C.P., Karimi, I.A., and Srinivasan, R. (2004). Novel solution approach for optimizing crude oil operations. *AIChE Journal* 50 (6): 1177–1197.

Seider, W.D., Seader, J.D., Lewin, D.R., and Widagdo, S. (2009). *Product and Process Design Principles*, 3e. New York: Wiley.

Shobrys, D.E. and White, D.C. (2000). Planning, scheduling and control systems: why can they not work together. *Computers & Chemical Engineering* 24 (2): 163–173.

Siemens (2021). The partnership between Siemens and Process Systems Enterprise (PSE). https://new.siemens.com/global/en/products/automation/industry-software/siemens-and-process-systems-enterprise.html (accessed 11 November 2021).

Speight, J.G. (2011). Chapter 2: Refining processes. In: *The Refinery of the Future*, 43. Boston: Elsevier.

Virtual Materials Group (VMG) (2021). Virtual materials process simulation. http://www.virtualmaterials.com/vmgsim.html (accessed 11 November 2021).

Williams, H.P. (1999). *Model Building in Mathematical Programming*, 4e. Chichester, West Sussex, England: Wiley.

2

Basic Petroleum Refinery Economics

2.1 Refinery Economics Overview

This section introduces basic concepts on refinery economics, which are useful to elucidate the materials and concepts employed throughout this book. A main component concerns refinery profitability and margins, as well as their trends and strategies for improvement. Table 2.1 defines several common financial terms for refineries.

2.1.1 Refinery Profitability

Numerous economic factors determine refinery profitability that generally include resource availability (e.g. equipment, personnel, funds, or money), resource use efficiency (e.g. using productive economic tools to operate a refinery in the best possible way, i.e. economically optimal), operating costs (e.g. personnel, materials, and taxes), and revenue generated (e.g. according to market conditions). The typical major factors to operate a refinery more profitably include the following:

- raw materials: type and amount to purchase (crude oil or more accurately, a mixture of them in the case of refineries);
- products: type and amount to produce;
- operations: optimal operating conditions to determine the best stream dispositions for each process.

In this regard, capital projects (e.g. construction of new refineries or plants) can offer the required future resources using appropriate economic tools to achieve profitability from revenue attained based on market situations and operating costs. On the whole, manufacturing optimization of a refinery is challenging due to its complexity because most alternatives involve small changes to current operations based on design (e.g. deciding to process extra crude oils or make more products). The situation augurs for a systematic approach by means of suitable mathematical optimization techniques and tools as early as during the synthesis and design stage, which is emphasized in and throughout this book. In this regard, capital projects resulting from the synthesis and design study conducted will give or indicate the

Model-Based Optimization for Petroleum Refinery Configuration Design, First Edition. Cheng Seong Khor.

Table 2.1 Common financial terms in refining business.

Term	Definition
Revenue	Payments received by a refinery
Expenses	Money spent by a refinery during operations comprising cash and noncash operating costs as well as fixed and variable costs
Income	Leftover money after deducting all costs from revenues comprising raw material costs, cash operating costs, noncash operating costs (e.g. depreciation), and taxes
Margin	Difference between product revenues and raw material costs
Cash flow	Money actually received or paid out in cash transactions, e.g. revenues, raw materials, cash operating costs, and taxes (excluding depreciation as noncash operating cost)
Capital	Money used by companies to invest in new or upgraded equipment or facilities as well as expand into new areas (businesses) and other major project types
Return on capital employed	Interest obtained by shareholders on their investments
Return on equity	Money earned which comprises all funds generated by a project or the company as a whole (subsequently converted to a percentage of capital employed)

right resources for the future (e.g. in the face of future competition) or any plausible situation (i.e. market outcome).

The main purpose of refinery economics (and its analysis) is to select the lowest-cost alternative(s) that can meet (i.e. supply) the product demand. We can use several economic concepts or tools to evaluate options and make operating decisions, e.g. marginal economics, capacity utilization economics, breakeven spread analysis, balances (in terms of material or mass), quality of barrel analysis, and cost of clean products analysis. For the purpose of this book, we focus on the first of such methods called marginal or incremental economics as described by refinery margins.

2.1.2 Refinery Margins

Margin generally refers to the difference between product revenues and raw material costs. It is effectively the money that a refinery possesses to cover all of its expenses. On a general note, margin is a key indicator in the manufacturing sector (which includes refined petroleum products) that is tracked in industrial journals. It is also monitored across different manufacturing site types, which make different product mix that leads to different revenue yields. Several variants of refinery margins exist – this chapter concentrates on refinery net margin or net income, which can be expressed in terms of the money per barrel of crude oil processed.

2.1.3 Refinery Margin Calculations

Several metrics are available to quantify refining profitability, which includes gross margin (see Figure 2.1) and net margin (see Figure 2.2). Gross margin is simply the difference between revenue from oil products sale and cost of purchasing the crude oils as raw materials. On the other hand, net margin is key toward determining refining profitability and can be described as the difference between revenue as given by the summation of product prices multiplied by their respective yields minus all the following associated costs:

- other feedstock cost, which includes that for purchasing additives (e.g. oxygenates for fuels production) and purchased gas condensate;
- variable cost, which includes processing requirements such as fuels, electricity, steam, catalysts, chemicals, and cooling water;
- fixed operating and maintenance costs, which include those involving personnel, supervisory, and clerical;
- engineering, maintenance materials, laboratories, property taxes, insurance and corporate overhead, and subcontractors;
- loss cost, which includes the cost of allowable and non-allowable hydrocarbon wastes produced during processing.

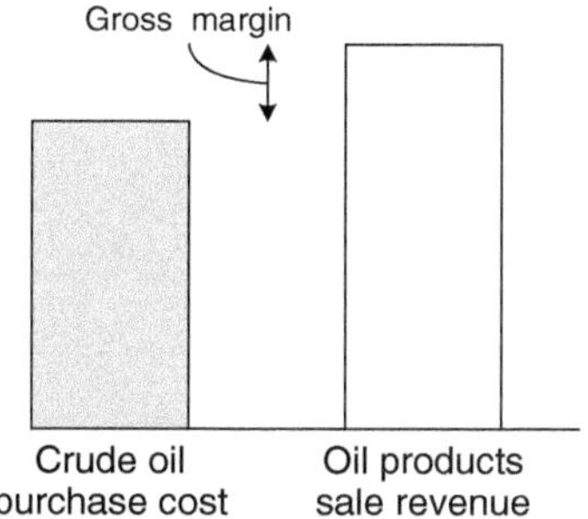

Figure 2.1 Gross margin.

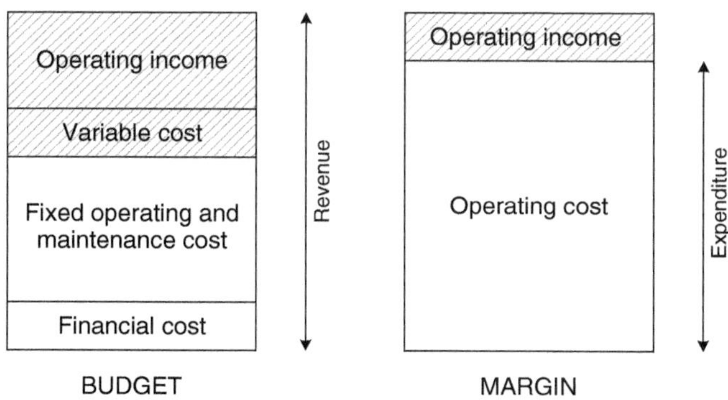

Figure 2.2 Refining net margin as given by the difference between budget and margin.

2.1.4 Refinery Margin Trends

Two significant considerations which affect the refining output profile and its margin (including return on capital) are the crude oil types processed (namely, light or heavy) and the refinery configuration. Standard setup examples of the latter include the following:

- simple (i.e. a hydroskimming refinery with a distillation unit and a reformer),
- semi-complex (i.e. with a visbreaker added to a hydroskimming refinery),
- complex (i.e. with a catalytic cracker, hydrocracker, or other advanced upgrading units).

The other considerations are supply and demand of crude oils and their associated products as regards political, technological, environmental, and governmental policies of countries and international assemblies or organizations. An additional consideration is the utilization rate of primary distillation capacity and secondary processing units (BP 2014).

2.1.5 Refinery Margin Improvement

Besides considering the mentioned significant factors on refinery profitability and company profile in general, refinery economics improvement can also consider the following factors:

- long-term refinery planning, e.g. through conventional linear program modeling and margin calculations;
- long-term investment planning, e.g. through master plans, feasibility studies, benchmarking, and management information systems;
- unit and overall refinery complex reliability;
- operational optimization (in real-time or online mode) of individual units;
- energy management and planning.

2.2 Marginal Economics for Incremental Optimization

Optimizing refinery operations is challenging and complex. To aid with the analysis of results or outcomes of formal mathematical optimization-based methods (e.g. mixed-integer superstructure optimization as presented in this book, which attempts to model and optimize an entire manufacturing site before and after a certain change is applied), techniques exist that consider small operational changes, e.g. feeding more crude oils for processing and manufacturing of more products. The concept of marginal economics underpins such techniques that can be less complicated and faster at identifying, eliciting, or predicting economically optimal decisions.

Marginal economics focuses on the outcome of incremental (i.e. marginal) changes from the base case instead of considering an entire manufacturing site. Such an approach often applies to a relatively feasible base case. To maximize profit, marginal economics identifies available options whose costs are less than expected revenue. Marginal changes or steps which result in costs equal to the expected revenue are called breakeven options. In this regard, steps resulting in higher costs than expected revenue are avoided.

Profit maximization can be attained by considering marginal or incremental revenues and costs. The approach can be defined as follows: incremental revenue is the slope of the revenue curve (which relates income to the amount of sold product); similarly, incremental cost is the slope of the cost curve, which relates expenditure to the amount (number) of manufactured products at a specified production level. A maximum profit can be estimated when incremental revenue is equal to incremental cost, as illustrated in Figure 2.3. At maximum profit, the next iteration gives a cost greater than revenue, which is not optimal and also not desirable because this situation means money is lost.

Other available tools to perform marginal economics analysis, which involves computing the value or cost (depending on which is applicable) of a last increment, include breakeven values, quality barrel costs, and cost of clean products (Maples 2000; Gary et al. 2007).

Marginal economics is applicable to study alternatives for site considerations or questions including feedstock value as determined by products' value that can be made from it (e.g. gasoline), cost of a certain product and its use to maximize profit, optimal operating conditions for a processing unit, and possible greater value of feedstock to another manufacturing site with different processing capabilities (or capacities). Marginal economics can also be used to assess whether investment in an equipment or project can generate enough revenue for an acceptable profit or rate of return.

Such economic analysis typically involves a stepwise, sequential procedure. Common steps for an economic evaluation of a manufacturing site involve the following: defining a base case of normal operating conditions, defining an alternative case of a proposed change to be considered, determining a net effect

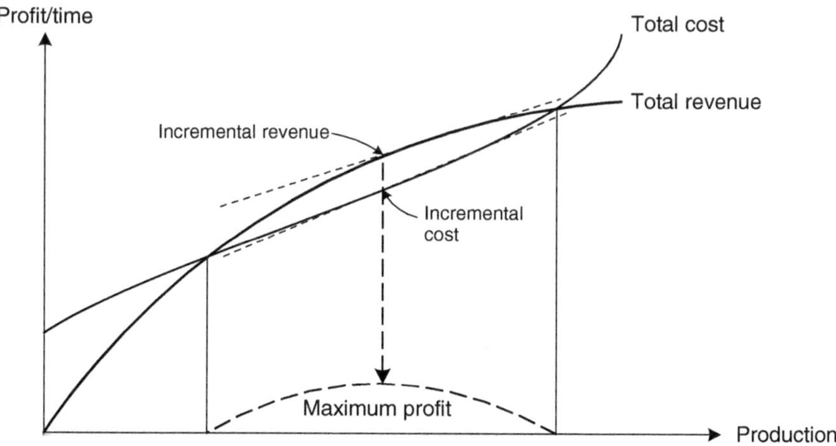

Figure 2.3 Basis for marginal economics to attain maximum profit when incremental revenue equals incremental cost.

Table 2.2 Common marginal economics terms in refining business.

Term	Definition
Cost	Resource (in monetary value) to be given up (willingly) in acquiring desired item
Price	Cost per unit of desired item
Value	Worth of desired item; crude oil value is evaluated based on its products (e.g. gasoline, kerosene, and diesel) which can be sold in the market
Incentive	Difference between value and cost (typically equal to profit)
Marginal value	Value assigned to an item in a marginal economic evaluation, also called shadow price or dual value
Mechanism	Physical steps to attain the desired result in marginal economics analysis
Relevant range	Validity range of marginal value attained, typically determined by operating limits (e.g. mechanism in operating a processing unit to produce more products changes when the unit is full (i.e. at limit) as is its product marginal value)

resulting from the considered change, performing a cost-benefit analysis of the proposed change, and finally deciding if the change is profitable. Table 2.2 define several common marginal economics terms for refineries.

2.3 Refinery Economic Analysis

Performing refinery economic evaluations follows stepwise procedures that are typical of the following:

1. *Develop a base case*: Define representation for normal operating conditions.
2. *Develop an alternative case*: Define consideration for changes in operating conditions.
3. *Develop a balance case (with intervention)*: Propose computation (i.e. model) to quantify the net effect that results from making change.
4. *Develop incentives*: Perform benefit and cost analysis to compare and contrast the proposed change.
5. *Develop decisions*: Determine whether making the proposed change is profitable.

2.3.1 Refinery Value Determination

Value determination is an exercise in which refineries assign values to raw materials, process streams, or products as realistically as possible by considering the following aspects:

- *Netback*: Also called site realization value that is the product's manufacturing site billing price or equivalently, the free-on-board price (i.e. the price paid by a buyer who is responsible for payment to get a product).

- *Replacement cost*: cost of a product that is the output of an alternative operation that has to be cut back by an amount equivalent to unsold additional product volumes because market demand is already satisfied.
- *Alternative use value*: Establish a material stream value in another product or as feedstock to a process that generates another product if there is no direct value for this stream.

2.3.2 Refinery Economic Evaluation

2.3.2.1 Simple Example

The following illustrates an application of the steps involved in performing economic evaluation for blending 1000 barrels per day (B/D) of naphtha into gasoline (Figure 2.4).

Step 1: Base Case. We begin with a base case that typically describes the current refinery operation (which is considerably optimized although not in a strict mathematical sense). The naphtha value for this disposition is assumed to be $70 per barrel.

Step 2: Alternative Case. An alternative case is the processing of naphtha in a reformer to produce reformate (which is a higher-quality gasoline blendstock valued at 78 $/barrel) and fuel gas (52 $/FOEB) (Note: FOEB stands for fuel oil equivalent barrel where a unit of FOEB equals 6.05 million of Btu (British thermal unit), i.e. 1 FOEB = 6.05 mmBtu). The reformer's operating costs are 900 $/day. Added to this case is butane at a rate of 100 barrels per day valued at 51 $/B. Note that since the reformate vapor pressure is lower than that of naphtha, butane addition to the gasoline blend does not violate vapor pressure specifications. The question to answer is: which option should be taken.

Step 3: Balance Case. This technique combines the available or gathered information on volumes, values, and revenues or costs in arriving at a determined net effect through the calculation for each output stream as given by the relation:

$$\text{Naphtha value} = \text{Output} - \text{Inputs} - \text{Operating costs}$$

Thus, the value of naphtha to reforming can be determined as 72.00 $ per barrel (which is computed by: −5100 $/D + 7800 $/D + 70,200 $/D − 900 $/D = 72,000 $/D ÷ 1000 barrel) (Table 2.3).

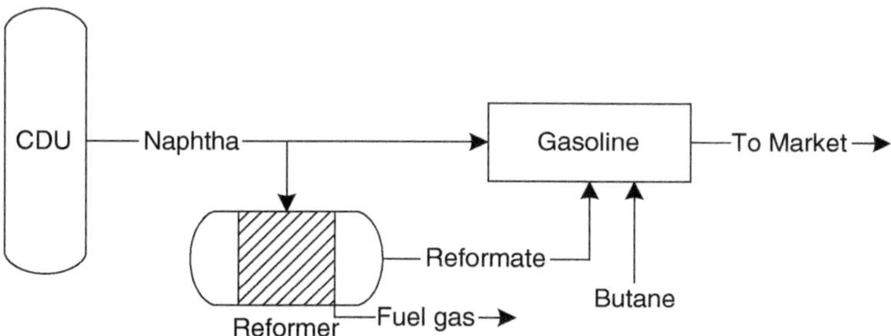

Figure 2.4 An application of refinery economic evaluation for blending naphtha into gasoline.

Table 2.3 Economic evaluation illustrative example of output value determination.

	Stream	Rate (B/D)	Cost ($/barrel)	Total daily value ($/day)
Input	Naphtha	−1000	x	−1000 x
	Butane	−100	51	−5100
Output	Fuel gas	150 (FOEB)	52 $/FOEB	7800
	Reformate	900	78	70,200
		Operating cost debit (cost increase)		(900)

Step 4: Benefit-Cost Analysis. This step compares the Alternative Case to the Base Case in which it is discerned that it is worthwhile to change the existing operation:

Naphtha value as gasoline blendstock (Base Case)	70.00 $/barrel
Naphtha value as reformer feed (Alternative Case)	72.00 $/barrel
Change incentive	+2.00 $/barrel gain
⇒ Decision: Send naphtha to reformer (for next processing)	

An imperative step is to ensure that all assumptions have been tested. It is important to consider other alternatives for testing (i.e. by posing the question: What else needs to be tested?) When we test the assumptions, it assists us in assessing how robust the decisions are.

It is also important to ask if there are other alternatives that should be evaluated. We can apply the principles of marginal economics to evaluate many alternatives. Doing so enables us to identify and implement the most plausible disposition for each stream at a manufacturing or refining site.

In essence, we can apply marginal economics to answer and determine various value determination exercises (exemplified in the following) in carrying out refinery economic evaluation:

- What a feedstock's value is based on the value of products that can be produced from it, and whether this feedstock gives greater value to another refining site with differing set of processing capabilities.
- What a particular product's cost is and where this stream should be utilized to gain maximum profit.
- What operating conditions are optimum for a processing unit.
- How (and why) investing in an equipment or a project will generate enough revenue for a profit or rate on return of investment that is adequate.

2.3.2.2 Advanced Example

PNT Limited is a large international oil company.

- KTH refinery, which is the smallest in Malaysia, manufactures four product groups, namely, light ends of fuel gas, gasoline or petrol, distillates, and residues.

- We wish to conduct an economic evaluation to determine additional crude oil amount (volume) that can be potentially processed by this refinery or manufacturing site.
- We gather the following information:
- Purpose of evaluation: Is processing another 10,000 B/D (barrel per day) of medium sour grade crude oil type called "Bonito" give a desirable incentive?
- Viewpoint of evaluation: KTH refinery manufacturing site of PNT.
- Method of evaluation: manufacturing economics

Step 1: Base Case. No additional crude oil volume processing.
Step 2: Alternative Case. Run additional 10,000 B/D (or 1592 m^3/D) of Bonto crude oil.

  - An acceptance or measurement criterion can be postulated as a minimum incentive/return of $1000 per day.
  - The required information mainly includes prices and volumes of the crude oil inputs and other inputs as well as products, besides the associated operating costs.
  - On the other hand, data that have not changed relative to the base case is not needed as we apply principles based on marginal economics.
  - In this regard, for example, only variable but not fixed operating costs are needed.

Step 3: Balance Case. We are given data prices in US dollars. A table is developed that sets up Balance Case to run Alternate Case that shows the net effects of running additional crude oil volumes on the final products and operating costs.

  - A minimum incentive required is determined that gives (or indicates) similar effects to operating costs.
  - The Balance Case is completed by computing the necessary values as tabulated in Table 2.4.

Thus, the value of Isthmus crude oil is determined as 69.30 $ per barrel (which is computed as: $(-104 + 284 + 228 + 104 - 26 - 1)$ k$/D = 693 k$/D ÷ 10 kB/D).

Table 2.4 Example 2.

	Stream	Rate (B/D)	Cost ($/B)	Total daily value (k$/day)
Inputs	Crude oil	(10,000)	x	
Outputs	Fuel gas (FOEB)	2000	52	104
	Gasoline	4000	71	284
	Distillate	3000	76	228
	Fuel oil	2000	52	104
Operating costs				(26)
Incentive required				(1)

2.3.2.3 Further Example

Example 3 illustrates a treatment of the preceding Example 2 using a volumetric unit of measure and Northwest Europe price data. Similarly, we develop a table that sets up Balance Case to run Alternate Case that shows the net effects of running additional crude oil volumes on the final products and operating costs. The resulting calculations and analysis are tabulated in Table 2.5. Thus, the value of Bonito crude oil is determined as 70.20 \$ per barrel (which is computed as $(-105+277+243+104-26-1)$ k\$/D = 702 k\$/D ÷ 10 kB/D).

The next scenario considers what the revenue reduction is if the fuel gas value decreases (e.g. by \$1 per barrel).

$$2000\ \text{FOEB/D Fuel Gas} \times \$1/\text{FOEB} = (2000)\ \$/\text{D}$$

$$(318\ \text{FOET/D} + 6.29\ \$/\text{FOET} = (2000)\ \$/\text{D}$$

The incentive would be: $+1000 - 2000 = -1000$ \$/D ⇒ implies a loss. The result indicates why it is imperative to test assumptions that are considered in a computational procedure to ascertain that the decision or outcome is robust.

To answer the question of whether we should run this Bonito crude oil if it is available at the calculated value, other factors that should be taken into consideration include market demand change as well as price and cost change. In addition, all of these factors are to be weighted appropriately to reflect their likelihood or probability of being realized. On the likelihood of whether product price would remain the same or increase as compared to the possibility that the price would decrease, there remains another possibility that the crude oil price could also change.

In retrospect, the foregoing analysis allows us to address the frequently asked questions: What is the maximum delivered price per barrel payable for a certain crude oil amount that still realizes the expected incentive which commensurate? If the particular crude oil is currently sold for the stated delivered price, what would the incentive be if the fuel gas price declined by a unit monetary amount (i.e. \$1) per barrel (or \$1/ton)? And ultimately, should the refinery procure and operate the said crude oil amount at the price previously determined? (Table 2.6).

Table 2.5 Example 3.

	Stream	**Rate (m^3/D)**	**Specific gravity**	**Cost (\$/ton)**	**Reference specific gravity**	**Total daily value (k\$/day)**
Inputs	Crude oil	1592	0.9350	x		
Outputs	Fuel gas (FOET)	318	1.000	330	(null)	105
	Gasoline	636	0.7400	576	0.7550	277
	Distillate	477	0.8350	603	0.8450	243
	Fuel oil	318	0.9910	330	(null)	104
Operating costs						(26)
Incentive required						(1)

Table 2.6 Summary of suggested procedure for refinery economic evaluation.

1. The total outputs (revenues) are summed.
2. The required (desired) incentive is subtracted from the revenues.
3. The total inputs (costs) are summed.
4. The operating costs are added to other costs.
5. The total costs are subtracted from the remaining revenue after deducting the incentive, which gives the total crude oil value. Finally, the total crude oil value is divided by the 1000 barrels of volume to express the value per barrel ($/B) of the crude oil.

2.3.3 Refinery Contracts

Refinery EPIC contracts cover activities referred to as engineering (E), procurement (P), fabrication, installation (I), and commissioning (C). Such agreements involve awarding a major project spanning detailed design until commissioning to a single contractor or a consortium of companies (i.e. as contractors). Contracts called "umbrella" or call-off agreements cover multiple assignments, which can be sequential or concurrent that are usually issued in the form of project orders. Umbrella contracts are established through competitive tenders with specified terms, schedules, and remuneration rates as adhering to company terms.

2.4 Concluding Remarks

This chapter provides an overview of relevant refinery economics concepts, metrics, and methodologies that can be suitably applied within a mixed-integer superstructure optimization approach (or framework as discussed in a later chapter) for refinery configuration synthesis and design. An introduction to basic refinery economics is presented, which covers refinery margin calculations as practiced in the industry besides prescribing some common strategies and variations for their possible improvement or enhancement particularly to perform incremental optimization. The next chapter discusses superstructure representations of topological alternatives for refinery configurations.

References

BP (2014). BP Statistical Review of World Energy 2014. London, UK.

Gary, J.H., Handwerk, G.E., and Kaiser, M.J. (2007). *Petroleum Refining: Technology and Economics*, 5e. New York: Marcel Dekker.

Maples, R.E. (2000). *Petroleum Refinery Process Economics*, 2e. Oklahoma: Pennwell.

3

Superstructure Representation

3.1 Introduction to Superstructures

An optimization approach in process synthesis mainly comprises three procedures: representation of design alternatives, mathematical modeling, and model solution. Many if not all of the possible flowsheet structures for a synthesis or design problem can be incorporated within a single representation, which gives rise to a diagram called a superstructure as shown in Figure 3.1. A superstructure embeds all feasible process design alternatives of interest by incorporating the different competitive process units and their possible interconnections via stream representations comprising inputs (e.g. feedstock or raw materials), intermediate materials, and outputs. Hence, each alternative can be a feasible or an optimal process flowsheet (Grossmann 1985).

3.2 Types of Superstructure Representation

There are three basic elements in a superstructure: states, tasks, and equipment as summarized in Table 3.1. States are a set of physical and chemical properties that identify a stream in a process, which can represent the feeds, intermediates, and final products – examples include temperature, pressure, mass flow, and composition of a stream. Tasks are the physical and chemical transformations of process operations that correspond to momentum, mass, and energy transfer operations such as reaction, absorption, and mixing. Task nodes represent the process operations that transform materials from one or more input states into one or more output states. Equipment are the physical devices that execute a given task such as reactor, heat exchanger, and distillation column (Yeomans and Grossmann 1999).

Three standard types of superstructure representation are state–task network (STN), state–equipment network (SEN), and resource–task network (RTN). More recent work has postulated alternative or modified superstructure representations with corresponding model formulations (Wu et al. 2016; Kelly et al. 2018).

Model-Based Optimization for Petroleum Refinery Configuration Design, First Edition. Cheng Seong Khor.

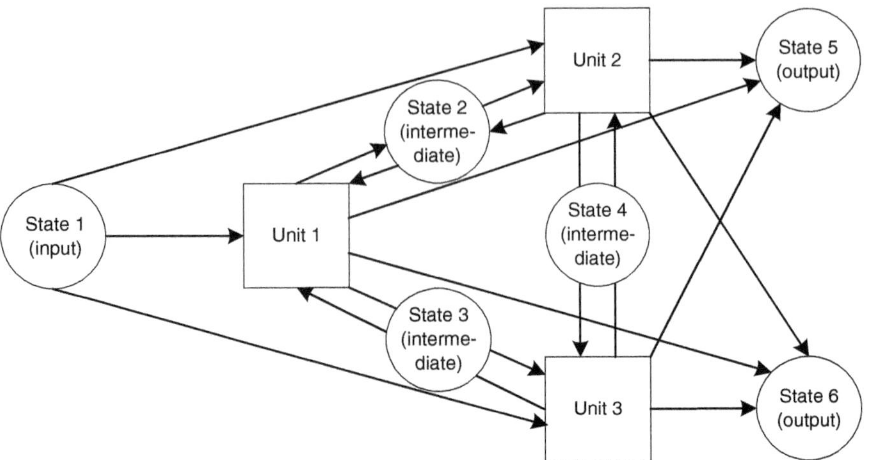

Figure 3.1 Illustration of a generic superstructure with multiple options of entities for selection.

Table 3.1 Basic superstructure elements.

Element	Description
States	Physical and chemical properties identity of process streams which represent feeds, intermediates, and final products
Tasks	Process operations that transform materials from input states into output states
Equipment	Physical devices that perform a task

3.3 State–Task Network Superstructure Representation

STN assumes that processing tasks produce and consume states (Figure 3.2). The states and tasks are defined first, leaving unknown equipment assignment to the second stage. Feedstock is connected to the product and vice versa via a path. Each

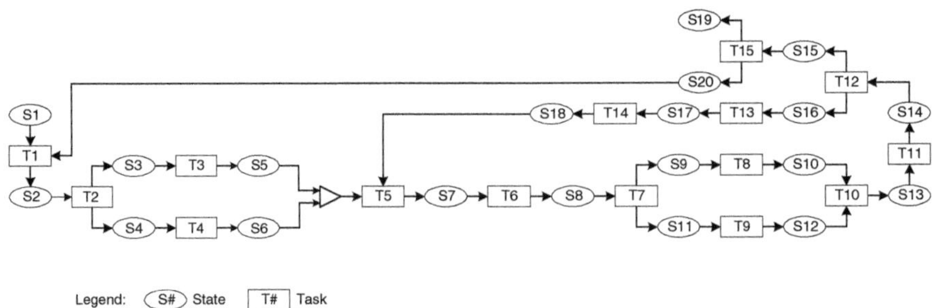

Figure 3.2 Example of state–task network (STN) superstructure representation for a process synthesis problem. Source: Adapted from Yeomans and Grossmann (1999).

intermediate state and task are on at least one such path. Some tasks are conditional; others must be present in all design alternatives. There is no need to distinguish one from the other at the level of representation but only at the level of modeling. One or more of these operations (temperature, momentum, mass, or energy transfer) can be performed in one task if technically feasible (Yeomans and Grossmann 1999).

When the states and the tasks are identified, the equipment assignment can be carried out in two ways: one-task–one-equipment (OTOE) assignment or the variable task-equipment (VTE) assignment. In OTOE, each task is distinguished from another and no two equipment are assigned the same task. If a task can be executed by two different equipment, the tasks are redefined to distinguish one from the other. Equipment assignment is explicitly performed, in which the case representation by both task and equipment is identical. The advantages of OTOE are that: (1) it is a straightforward representation from which an optimization model can be conveniently formulated; (2) it handles the assignment of equipment implicitly by reducing the combinatorial complexity of the mathematical model.

On the other hand, VTE enables a single equipment to be assigned different tasks and vice versa. A set of equipment that can perform all the tasks needed is first identified. Assignment of equipment to the task is part of the optimization decisions. A single equipment unit can be assigned to different tasks, and a single task can be assigned to different equipment. It is necessary to select one and only one of the equipment configurations available for the task (Yeomans and Grossmann 1999).

3.4 State–Equipment Network Superstructure Representation

The SEN superstructure representation considers full connectivity among states and equipment (Figure 3.3). It defines the tasks and equipment, leaving the assignment of tasks to equipment to be determined. It includes different states of process and equipment that are likely to be used. One or more different tasks which can be

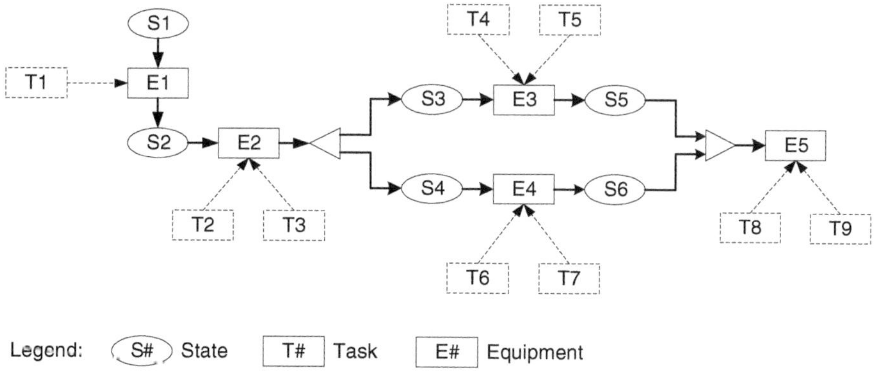

Figure 3.3 Example of state–equipment network (SEN) superstructure representation for a process synthesis problem. Source: Adapted from Yeomans and Grossmann (1999).

performed by single equipment are determined by considering full connectivity between states and equipment. Tasks that can take place on specific equipment are not prespecified. Assignment of task to equipment is part of the optimization model (task assignment is unknown), which is equivalent to the VTE variant of STN. SEN leads to a smaller combinatorial problem for equipment selection but implicit combinatorial complexity is present in the possible equipment interconnections. Also, state definition is not unique and all possible realizations of the streams that will originate from a certain task will have to be taken into account. This can complicate the modeling stage. The SEN representation is also useful for retrofit design problems as it shows explicitly the existing equipment (Yeomans and Grossmann 1999).

3.5 Resource–Task Network Superstructure Representation

The RTN superstructure representation assumes that a task only consumes and produces resources (Figure 3.4). The concept of resource includes all entities involved in the process steps such as materials (e.g. raw materials, intermediates, and products), processing and storage equipment (e.g. tanks and reactors), utilities (e.g. operators and steam), and equipment conditions (clean and dirty). There is no distinction between equipment of any type and other resources. All resources are allowed to be produced or consumed by the tasks at any time during their execution. This means that processing items are treated as though it is consumed at the start of a task and produced at the end. RTN is able to capture additional features of a problem straightforwardly, thus providing conceptual simplicity and direct applicability to a large number of complex problems such as process scheduling (Barbosa-Póvoa and Pantelides 1997, 1999).

Alternatively, a resource–technology network (also abbreviated as RTN) superstructure representation is proposed as a more generalized variant to the foregoing

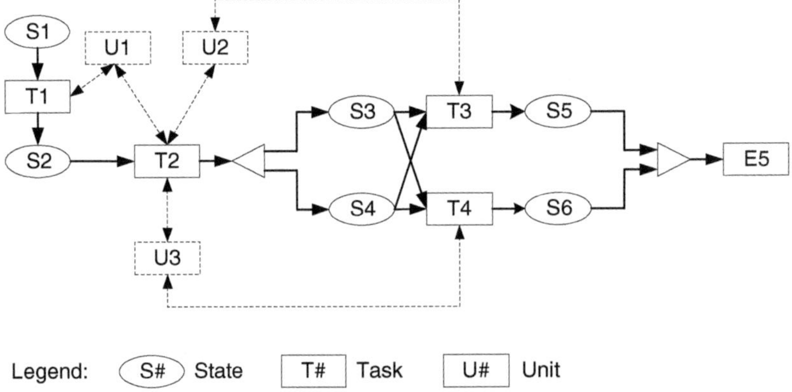

Figure 3.4 Example of resource–task network (RTN) superstructure representation for a process synthesis problem. Source: Adapted from Pantelides (1994).

Table 3.2 Comparison across superstructure representations.

	STN	SEN	RTN
Main approach	Prespecify task to consume and produce materials by using equipment and utilities	Assign one or more different tasks to equipment (not prespecified; fully connected)	Assign task to consume and produce resources (more generalized than STN/SEN)
Representation of state/task/equipment/resource	Restricted: each operation use only one major equipment item	No unique state definition	Include all resource types: no distinction between equipment type and other resources
Complexity	Simple to add number of units and interconnections; reduce search alternatives	Stream properties and quality (properties) depend on selected task for equipment	Straightforward representation; applicable to many complex problems

Source: Adapted from Pantelides (1994).

RTN in that it allows resource storage along the process steps (or pathways), incorporates technological modes, and caters for various alternative network pathways leading to the same output state. This superstructure type is initially postulated to model urban energy systems and subsequently applied by Samsatli et al. (2015) to whole system biomass value chain analysis and optimization.

A comparison across the variants of superstructure representations is given in Table 3.2. In this work, we construct a superstructure using STN representation, specifically the OTOE variant.

3.6 Superstructure Generation

Synthesis and design activities in petroleum-based processes for refining and petrochemical production (i.e. manufacturing in general) are complex involving multiple products with low (even very little) similarities. Thus, efforts to develop a generic problem representation such as manufacturing recipes (i.e. in consequence of or as informed by, for example, an optimal production plan or schedule) might need apparently elaborate interventions or definitions on the part of a modeler or user.

With these challenges in mind, procedures to generate a representative superstructure have been postulated to capture or embody the full complexity of a process synthesis problem without giving rise to ambiguity. At the same time, we desire for such techniques to incorporate elements or features in such a way that they strive toward achieving uniformity and minimizing user-dependent customizations (Yeomans and Grossmann 1999; Grossmann et al. 2000). In applying mathematical programming techniques to synthesis or design problems, a requisite initial activity is to postulate a superstructure of alternatives. This starting step is applicable to whether formulating a model that is aggregated (i.e. with low granularity) or more detailed (high granularity).

3.7 Other Superstructure Representations

3.7.1 State–Space Network Superstructure Representation

A state–space superstructure representation is devised by Bagajewicz et al. applied to process synthesis problems to perform mass or heat exchanger network design (Bagajewicz and Manousiouthakis 1992; Bagajewicz et al. 1998). Its use can be flexibly combined to interact with a few types of mathematical operators (mainly linear) to assist with solving problems of significant size. The state–space conceptual framework is comprehensive in allowing all possible design alternatives for a synthesis problem to be generated and represented. Based on dynamic systems theory, the associated superstructure representation as shown in Figure 3.5 (in which the level of detail can be aggregated or rigorous) treats a process as a distribution network in determining all the flows (which are handled by an operator).

3.7.2 Unit Operation–Port–State Superstructure Representation

A superstructure representation called unit-operation–port–state superstructure (UOPSS) is proposed by Kelly and coworkers, particularly for advanced modeling of planning and scheduling systems (Zyngier and Kelly 2012; Kelly et al. 2018; Franzoi et al. 2020) such as those typically encountered in refineries. This superstructure is developed by extending the associated process flow diagram (PFD) and/or piping and instrumentation diagram (P&ID) for a process or production plan to capture the procedures involved. This representation type is made available in the IMPL (Industrial Modeling and Programming Language) software package (Figure 3.6) (Kelly and Menezes 2019).

3.7.3 Bond Graph Superstructure Representation

Bond graph (including its generalized version) is proposed as a structured process modeling method with its foundation rooted in thermodynamics and network

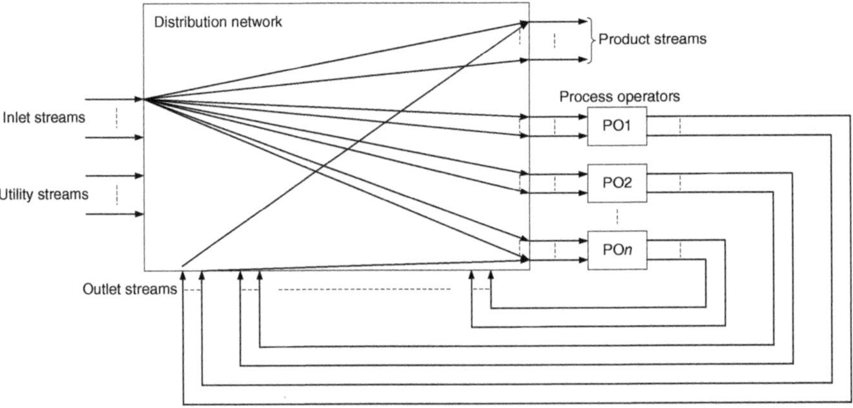

Figure 3.5 Example of state–space network superstructure representation for a process synthesis problem. Source: Adapted from Bagajewicz et al. (1998).

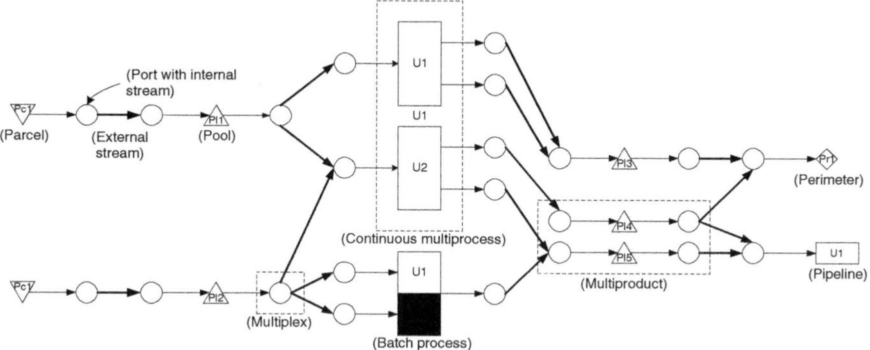

Figure 3.6 Example of unit operation–port–state superstructure (UOPSS) representation for an operation (production) problem. Source: Based on Kelly (2005) and Zyngier and Kelly (2012).

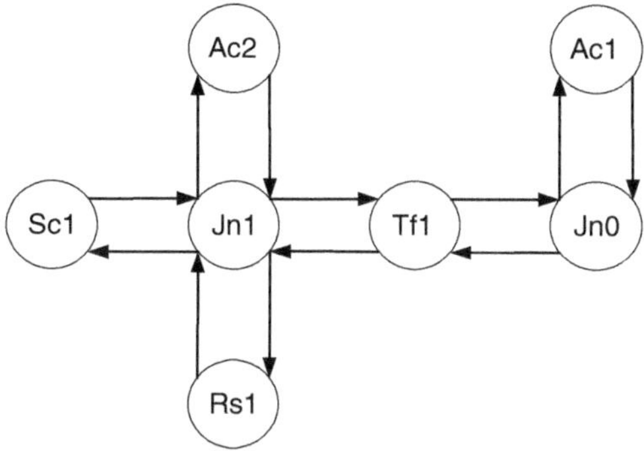

Figure 3.7 Example of bond graph superstructure representation. Note: Port types are denoted as follows: Sc1 = source; Tf1 = energy transformer; Rs1 = resistance/dissipation; Ac1, Ac2 = accumulation; Jn0, Jn1 = connection (0-junction [parallel] and 1-junction [series], respectively).

theory, and its representation made up of multi-port elements (Couenne et al. 2008; Borutzky 2010). The approach has been applied to several chemical process applications including reactor network modeling (Couenne et al. 2006) and fault detection and diagnosis (El Harabi et al. 2014, 2015; Sayed-Mouchaweh 2018). Figure 3.7 shows an example of a superstructure based on the bond graph modeling methodology (Díaz-Zuccarini and Pichardo-Almarza 2011).

3.8 Superstructure Representation Example for Naphtha Processing

We develop a STN-based superstructure representation that is sufficiently rich to embed all possible alternative topologies for the subsystem of naphtha produced from the atmospheric distillation unit (ADU) of a refinery as shown in Figure 3.8. The first refining processing step is crude distillation wherein crude oil (CR) is distilled into oil fractions with respect to its boiling points. Naphtha constitutes the

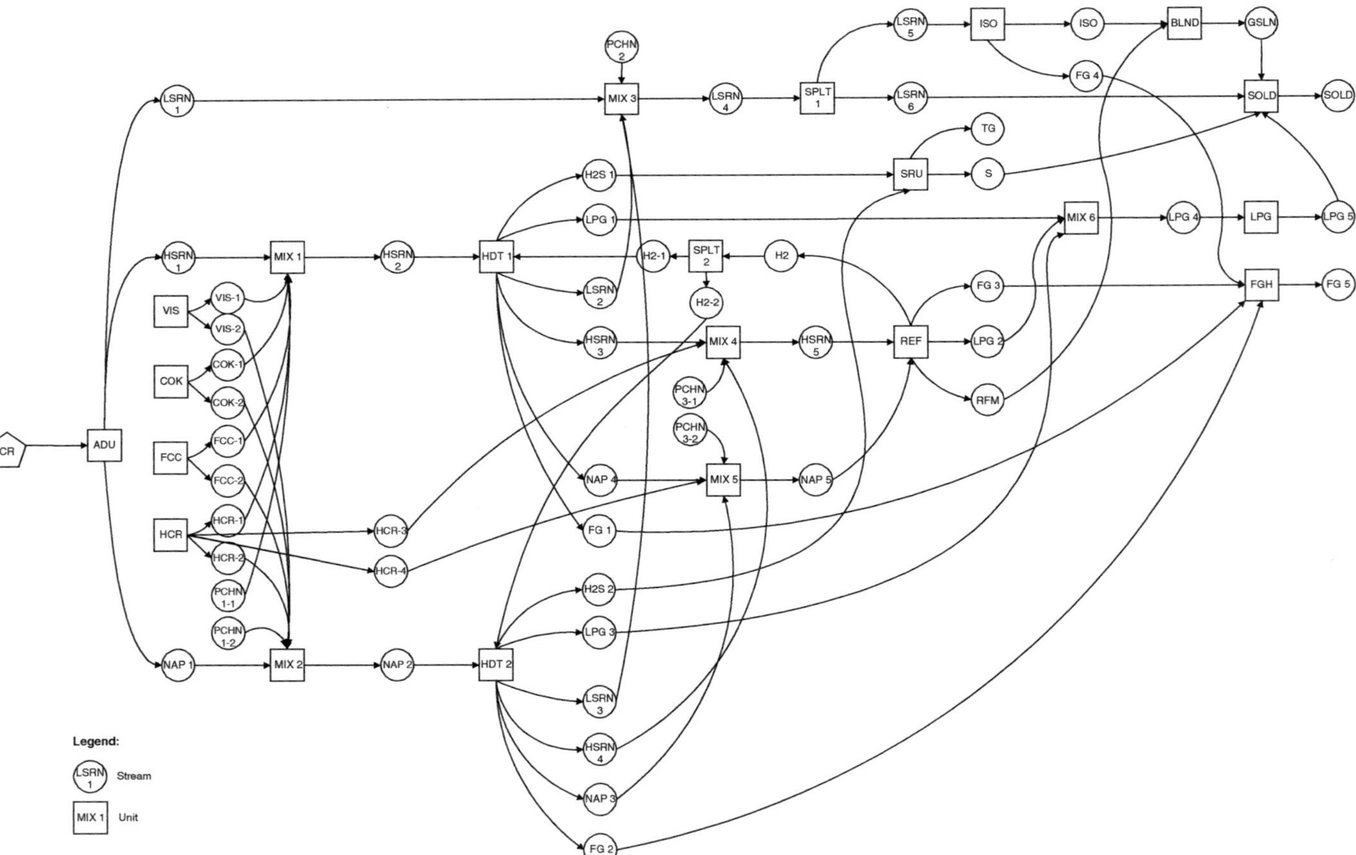

Figure 3.8 State–task network (STN) superstructure representation for naphtha produced from atmospheric distillation.

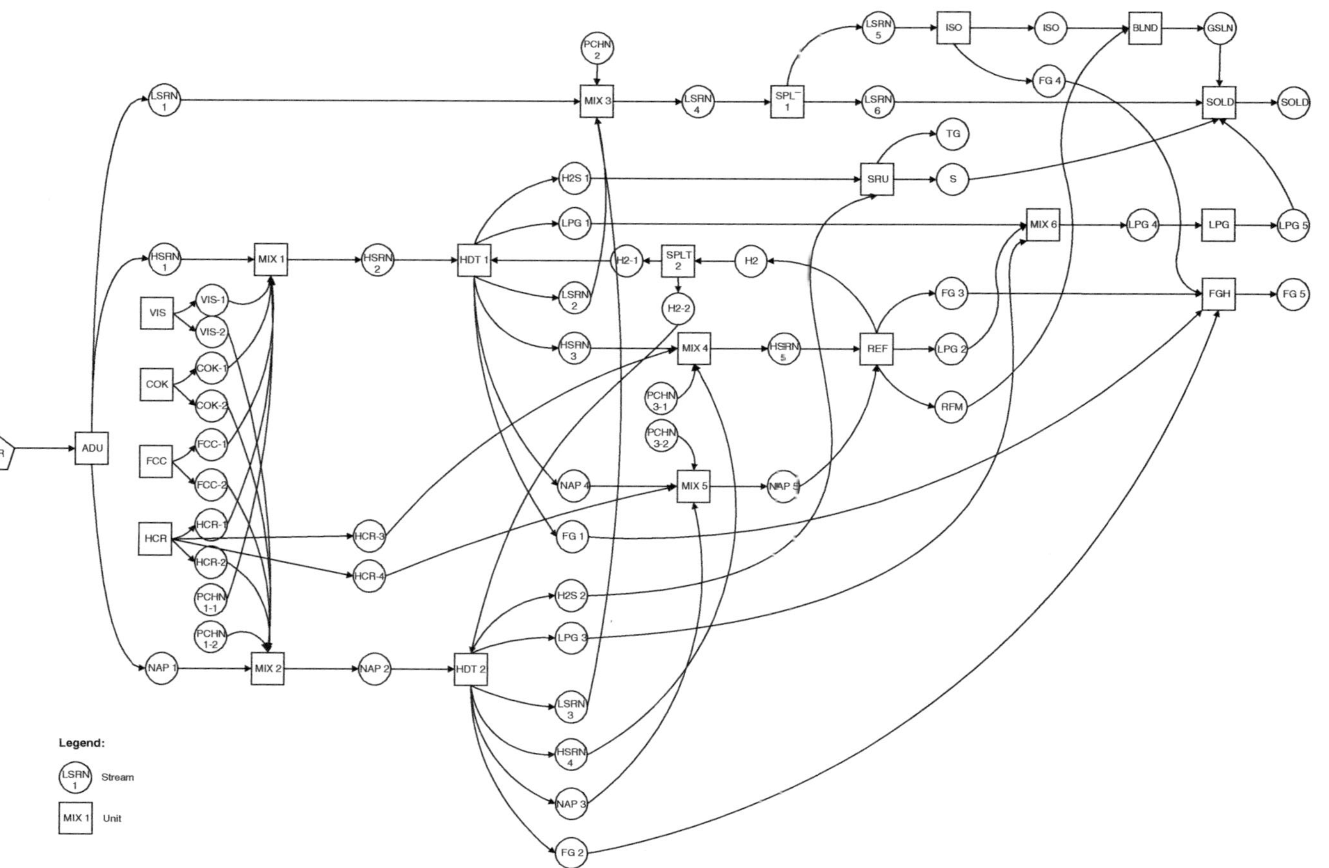

Figure 3.8 *(Continued)*

Abbreviations:

CR	Crude oil	HDT	Hydrotreater
ADU	Atmospheric distillation unit	LPG	Liquefied petroleum gas
LSRN	Light straight run naphtha	H2	Hydrogen
HSRN	Heavy straight-run naphtha	ISO	Isomerization unit
NAP	Naphtha	SRU	Sulfur recovery unit
MIX	Mixer	REF	Reformer
SPLT	Splitter	S	Sulfur
VIS	Visbreaker	FG	Fuel gas
COK	Coker	BLND	Blending
FCC	Fluidized catalytic cracker	FGH	Fuel gas header
HCR	Hydrocracker	GSLN	Gasoline
PCHN	Purchased naphtha	TG	Tail gas

Figure 3.8 (*Continued*)

lighter fractions that are obtained from this preliminary process. Depending on the distillation column design as well as the refinery economics, the ADU can produce light straight-run naphtha (LSRN1) and heavy straight-run naphtha (HSRN1) or an undifferentiated class of naphtha called wild naphtha (NAP1). In the first case, LSRN1 is sent to a mixer (MIX3) together with purchased naphtha (PCHN2) and LSRN2 from the hydrotreater (HDT1). The output from MIX3, which is LSRN4, has two processing routes: as feedstock for isomerization unit (ISO) or as a final product to be sold.

Isomerization yields isomerate (ISO), which is a blending component for gasoline (GSLN). Meanwhile, HSRN1 is mixed with naphtha from the cracking of heavier fractions in MIX1 before it is sent to HDT1 to be desulfurized.

HDT1 yields hydrogen sulfide gas (H2S1), liquefied petroleum gas (LPG1), desulfurized naphtha, and fuel gas (FG1). H2S1 is sent to the sulfur recovery unit (SRU) where sulfur (S) is extracted and finally sold.

All LPG product streams (LPG 1, LPG 2, and LPG 3) are sent to MIX6 and subsequently to LPG recovery unit (LPG) from which treated LPG (LPG 5) is sold. Similar to the output from the distillation unit (ADU), the desulfurized naphtha from HDT 1 can be classified as light (LSRN 2) and heavy (HSRN 3) *or* undifferentiated (NAP 4). LSRN 2 is mixed with LSRN 1 and PCHN 2 in MIX 3, as stated above.

On the other hand, HSRN3 is sent to a mixer (MIX4), possibly with purchased naphtha (PCHN3_1) and/or naphtha from the hydrocracker (HCR). The output of MIX4 (i.e. HSRN5) is the feedstock for the reformer (REF). FG1 goes to the fuel gas header (FGH), which supplies fuel gas (FG5) to the entire refinery. In the case that undifferentiated naphtha (NAP4) is produced from HDT1, it will also be mixed with purchased naphtha (PCHN3_2) *and/or* naphtha from the hydrocracker (HCR) in MIX5, whose output (NAP5) is sent to REF. The products from REF are hydrogen gas (H2), fuel gas (FG3), liquefied petroleum gas (LPG2), and reformate (RFM). H2 is fed to HDT1, while RFM is used as a gasoline blending component. FG3 is sent to FGH.

For the case of undifferentiated naphtha (NAP1) produced from ADU, the processing route is similar in that NAP1 is mixed with naphtha from cracking processes in MIX2 before being hydrotreated in HDT2. The products from HDT2 are H2S2, LPG3, desulfurized naphtha, and FG2. Each product has the exact same route as the products from HDT1.

Other than distillation, naphtha is also produced from the cracking of distillation bottoms in the visbreaker (VIS), catalytic cracker (FCC), hydrocracker (HCR), or coker (COK). Visbreaker has the lowest severity, while coker has the highest. The intermediate products from VIS, COK, FCC, and HCR are assumed to be heavy naphtha with API ≤33° for medium or heavy crudes and >33° for light crudes.

The processing of medium and heavy crude requires more severe processes, hence COK, FCC, and HCR are enforced to exist for the associated constraint stipulated. On the other hand, light crude processing requires less severity, hence only VIS and FCC are enforced to exist.

3.9 Chapter Summary

Three basic superstructure representations are STN, SEN, and RTN. Each representation has certain features (and limitations) that can be applied and exploited to cater for specific problem requirements in exploiting its suitability and advantages. The beneficial features mainly show impact in terms of the resulting computational efficacy and efficiency in converging toward an acceptable (if not optimal or near-optimal) solution. Other superstructure types have been reported also in the literature (and introduced in this chapter) but the aforementioned three basic types remain widely used particularly for process synthesis problems.

References

Bagajewicz, M.J. and Manousiouthakis, V. (1992). Mass/Heat-exchange network representation of distillation networks. *AIChE Journal* 38: 1769.

Bagajewicz, M.J., Pham, R., and Manousiouthakis, V. (1998). On the state space approach to mass/heat exchanger network design**First presented in the 1990 Annual AIChE Meeting in Chicago, paper #22d. *Chemical Engineering Science* 53 (14): 2595–2621.

Barbosa-Póvoa, A.P.F.D. and Pantelides, C.C. (1997). Design of multipurpose plants using the resource-task network unified framework. *Computers & Chemical Engineering* 21: S703–S708.

Barbosa-Póvoa, A.P.F.D. and Pantelides, C.C. (1999). Design of multipurpose production facilities: a RTN decomposition-based algorithm. *Computers & Chemical Engineering* 23: S7–S10.

Borutzky, W. (2010). Bond graph modelling of open thermodynamic systems. In: *Bond Graph Methodology: Development and Analysis of Multidisciplinary Dynamic System Models* (ed. W. Borutzky), 425–468. London: Springer London.

Couenne, F., Jallut, C., Maschke, B. et al. (2006). Bond graph modelling for chemical reactors. *Mathematical and Computer Modelling of Dynamical Systems* 12 (2–3): 159–174.

Couenne, F., Jallut, C., Maschke, B. et al. (2008). Structured modeling for processes: a thermodynamical network theory. *Computers & Chemical Engineering* 32 (6): 1120–1134.

Díaz-Zuccarini, V. and Pichardo-Almarza, C. (2011). On the formalization of multi-scale and multi-science processes for integrative biology. *Interface Focus* 1 (3): 426–437.

El Harabi, R., Ould-Bouamama, B., and Abdelkrim, M.N. (2014). Bond graph modeling for fault diagnosis: the continuous stirred tank reactor case study. *SIMULATION* 90 (4): 405–424.

El Harabi, R., Smaili, R., and Abdelkrim, M.N. (2015). Fault diagnosis algorithms by combining structural graphs and PCA approaches for chemical processes. In: *Chaos Modeling and Control Systems Design* (ed. A.T. Azar and S. Vaidyanathan), 393–416. Cham: Springer International Publishing.

Franzoi, R.E., Menezes, B.C., Kelly, J.D. et al. (2020). Cutpoint temperature surrogate modeling for distillation yields and properties. *Industrial & Engineering Chemistry Research* 59 (41): 18616–18628.

Grossmann, I.E. (1985). Mixed-integer programming approach for the synthesis of integrated process flowsheets. *Computers and Chemical Engineering* 9 (5): 463–482.

Grossmann, I.E., Caballero, J.A., and Yeomans, H. (2000). Advances in mathematical programming for the synthesis of process systems. *Latin American Applied Research* 30 (4): 263–284.

Kelly, J.D. (2005). The unit-operation-stock superstructure (UOSS) and the quantity-logic-quality paradigm (QLQP) for production scheduling in the process industries. G. Kendall, L. Lei, and M. Pinedo. *Multidisciplinary Conference on Scheduling Theory and Applications (MISTA)*, New York City, United States (18–21 July 2005), INFORMS (The Institute for Operations Research and the Management Sciences).

Kelly, J.D. and Menezes, B.C. (2019). Industrial modeling and programming language (IMPL) for off- and on-line optimization and estimation applications. In: *Optimization in Large Scale Problems: Industry 4.0 and Society 5.0 Applications* (ed. M. Fathi, M. Khakifirooz, and P.M. Pardalos), 75–96. Cham: Springer International Publishing.

Kelly, J.D., Menezes, B.C., and Grossmann, I.E. (2018). Successive LP approximation for nonconvex blending in MILP scheduling optimization using factors for qualities in the process industry. *Industrial & Engineering Chemistry Research* 57 (32): 11076–11093.

Pantelides, C.C. (1994). Unified frameworks for optimal process planning and scheduling. In: *Second International Conference of the Foundations of Computer-Aided Process Operations (FOCAPO)* (19–24 June 1993) (ed. D.W.T. Rippin, J. Hale, and J.F. Davis), 253–274. Crested Butte, Colorado, USA: CACHE Corporation.

Samsatli, S., Samsatli, N.J., and Shah, N. (2015). BVCM: a comprehensive and flexible toolkit for whole system biomass value chain analysis and optimisation – mathematical formulation. *Applied Energy* 147: 131–160.

Sayed-Mouchaweh, M. (2018). *Fault Diagnosis of Hybrid Dynamic and Complex Systems*. Cham: Springer International Publishing.

Wu, W., Henao, C.A., and Maravelias, C.T. (2016). A superstructure representation, generation, and modeling framework for chemical process synthesis. *AIChE Journal* 62 (9): 3199–3214.

Yeomans, H. and Grossmann, I.E. (1999). A systematic modeling framework of superstructure optimization in process synthesis. *Computers & Chemical Engineering* 23: 709–731.

Zyngier, D. and Kelly, J.D. (2012). UOPSS: a new paradigm for modeling production planning & scheduling systems. In: *22nd European Symposium on Computer Aided Process Engineering* (ed. I.D.L. Bogle and M. Fairweather), 20. London: Elsevier.

4

Modeling Framework

This chapter presents a generic modeling framework of superstructure-based optimization using mixed-integer programming.

4.1 Modeling of Mixed Continuous and Integer Decision Variables

Discrete decisions are potentially useful and can constitute an important role in process design particularly synthesis activities to address conceptual process design problems. Typical examples of such discrete or integral decisions include integrating multiple process units in a flowsheet to model a process network, considering available reactor sizes, and representing the number of trays in a distillation column.

Optimization modeling techniques or mathematical programming approaches can use discrete or integer variables particularly zero–one (0–1) binary variables to represent these decisions. A value of one means the selection (i.e. existence or operation) of a unit or assignment of a task while a value of zero means the non-selection (i.e. absence or non-operation including possible shutdown) of a unit or task. By combining such discrete and continuous variables, conceptual process design problems can be modeled through an objective function and a set of constraints representing material balances and energy balances (the latter are particularly important to incorporate heat integration concepts) as well as design and structural specifications.

A nonlinear process network problem necessitates simultaneous optimization of not only conventional continuous variables (chiefly on or related to unit operating conditions and material stream qualities and quantities) but also that of the integer variables associated with discrete decisions (as mentioned earlier). As a result, the problem lends itself naturally to be formulated most generally as a mixed-integer nonlinear programming (MINLP) model. A basic formulation core of such a model comprises discrete decisions of structure to determine the flowsheet topology or shape and continuous decisions of design attributes to evaluate the flowsheet performance (Biegler et al. 1997; Grossmann 2002; Trespalacios and Grossmann 2014; Chen and Grossmann 2017).

Model-Based Optimization for Petroleum Refinery Configuration Design, First Edition. Cheng Seong Khor.

4.2 Superstructure Optimization Modeling

An optimization model aims to determine the optimal objective function value, typically one in terms of maximum profit, minimum cost, or minimum environmental impact (or a combination of these) by computing values of the variables, which satisfy equality and inequality constraints, thus yielding an optimal solution. The entailed model formulations mainly consist of constant-yield-based linear material balances and logical constraints enforcing the associated relevant design specifications and structural specifications. The latter stipulate all possible interconnectivity relationships among the process units and material streams for the selection and sequencing of alternative processing routes based on a suitable superstructure representation. Developing models to account for these constraints impacts and influences the ensued resultant solutions.

4.3 Constructing Superstructures

Basic elements for constructing a flowsheet superstructure chiefly involve process streams (including raw material feeds, intermediates, and products) and process units as illustrated in Figure 4.1. There are two general techniques to construct a flowsheet superstructure mainly intended for process synthesis activity:

- superstructure-based optimization as proposed by Sargent and Gaminibandara (1976) covers several techniques to construct superstructures, which include approaches developed for overall flowsheet optimization (Kocis and Grossmann 1989) and heat exchanger networks synthesis Floudas et al. (1986);
- graph theory-based method by Friedler et al. (1993).

Several other methods are available, and they have been reported in the literature that involve constructing superstructure in an ad-hoc manner (Bagajewicz and Manousiouthakis 1992; Bagajewicz et al. 1998; Dowling and Biegler 2015; Wu et al. 2016; Ramapriya et al. 2018). Such a wide range of approaches indicates that a suitable way to develop a superstructure is largely problem-dependent or application-specific (Chen and Grossmann 2017) (Figure 4.2).

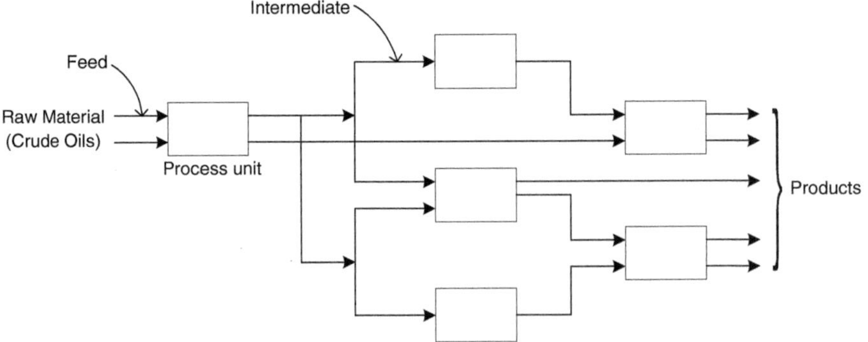

Figure 4.1 A generic superstructure construction for modeling a network of processes and materials of a typical oil refinery operation.

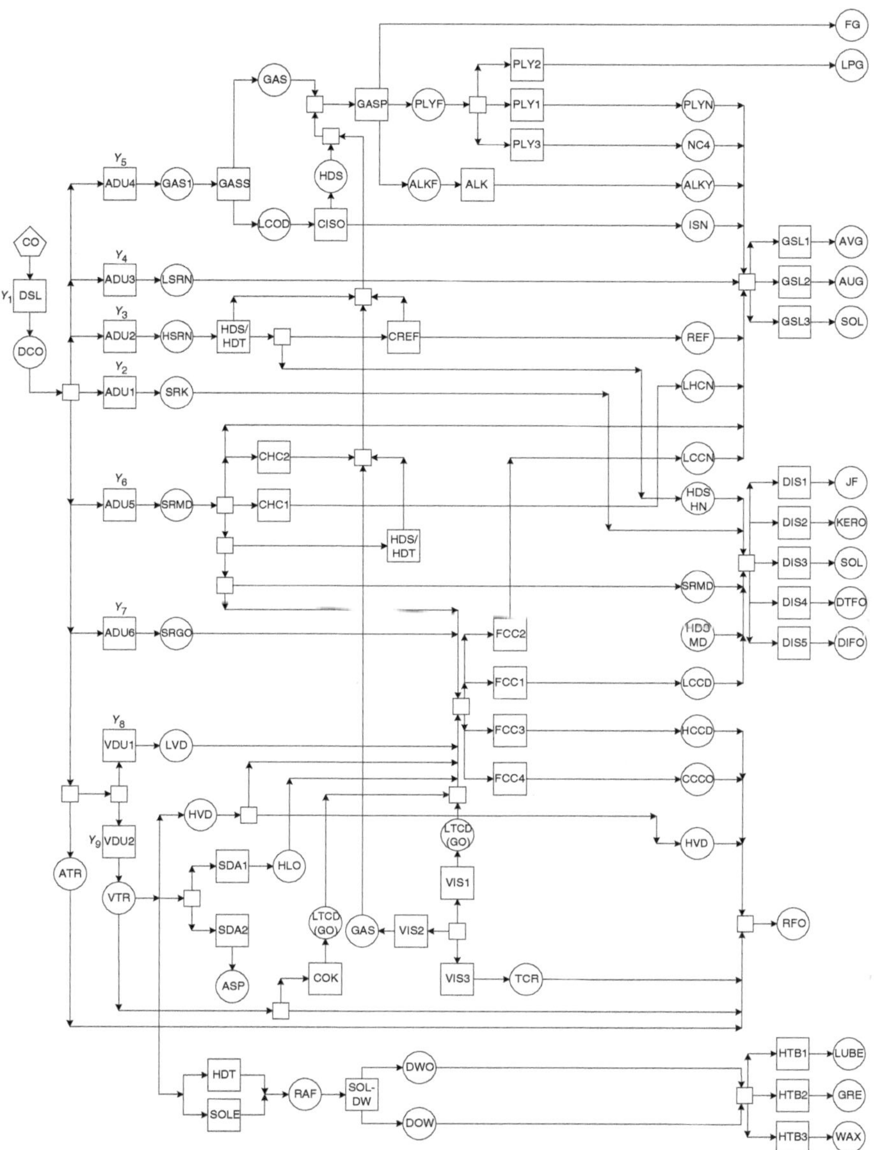

Figure 4.2 A generic superstructure construction for refinery operations (based on state–task network representation with one-task–one-equipment assignment).

4.4 Modeling of Superstructure Representations

Mathematical programming can accommodate synthesis and design models of various degrees of complexity. Addressing such complexities can be classified into the following three categories (see Table 4.1) based on increasing superstructure representation details and their corresponding model formulations (Yeomans and Grossmann 1999):

Table 4.1 Categories of superstructure representation in terms of detail or complexity.

Model type	Superstructure representation	Superstructure detail	Model formulation
Aggregate	Mostly linear	Low	Simplified with one dominant aspect (as objective)
Shortcut	Relatively simple nonlinear	Moderate	Cost optimization with simple unit performance
Rigorous	Highly nonlinear and possibly mixed-integer	High	Rigorous and complex

1) *Aggregate models (mostly linear)*: high-level superstructures with simplified formulations in terms of a dominant aspect or objective;
2) *Shortcut models (relatively simple nonlinear)*: moderately detailed superstructures with formulations that involve cost optimization but in which the unit performance is predicted with relatively simple nonlinear models to reduce the computational expense or exploit the algebraic structure to attain global optimality.
3) *Rigorous models (highly nonlinear and possibly mixed integer)*: detailed superstructures with rigorous and complex formulations to predict and prescribe unit performances (Table 4.1).

Aggregate models and superstructure representations yield simpler optimization model types of linear programs (LPs), non-LPs (NLP), or MILPs. They are easier to solve as compared to mixed-integer NLPs (MINLPs), which typically result from shortcuts and rigorous representations. The work discussed in this book largely adopts an aggregate modeling approach leading to MILP model formulation.

4.5 Modeling of Discrete Decisions and Logical Relations

A particular set of qualitative and quantitative information is usually available or can be specified at the start of process synthesis (or design in general) activities. In the context of petroleum refining or chemical process synthesis, such quantitative information can be in the form of rate expressions for reactions, thermodynamic property correlations, or sizing equations of various process unit models. On the other hand, synthesis-related qualitative information can be in the form of heuristics, logic of relations among the process units or streams (and their dispositions), and prior knowledge of typical or probable decisions leading to an acceptable design (or its promising alternatives for further evaluation).

Such qualitative information espoused potentially provides information about the decision space (feasible region) to explore for the best (i.e. optimal) design, thus resulting in a more defined problem (or tighter design decision space) that is easier to solve. Therefore, a suitable modeling and optimization framework has to be able to incorporate both quantitative relations in the form of equations (including inequalities) and qualitative (logical) information in a synthesis or conceptual design problem. Simply speaking, we must use an approach that is capable to handle both kinds of information within a common framework.

It is imperative to be able to express qualitative knowledge mathematically in integrating these two types of process-related information. Heuristics and logic relations, which describe connections and interactions among the process units in a superstructure, can be expressed as propositional logics and represented correspondingly as linear inequalities by using binary 0–1 variables (Raman and Grossmann 1991). Based on this form of mathematical expression to describe the process network representation for a developed superstructure, we can integrate heuristics knowledge and logic relations by using a quantitative framework as a basis to formulate the associated synthesis problem (Raman and Grossmann 1992, 1993a; Hooker et al. 1994). An example of the form for such a basis that is suitable for process synthesis or design includes an algorithmic approach like generalized disjunctive programming (GDP) (Turkay and Grossmann 1996; Trespalacios and Grossmann 2014, 2016) or a graph theoretic approach like P-graph (Friedler et al. 1993; Kovács et al. 2000). In turn, these forms of logic-based modeling methods can be used to derive mathematical programming (i.e. optimization) models (e.g. MINLP), which are composed of algebraic equations.

It is noteworthy that GDP or disjunctive programming in general is an efficient formulation with a structure that is well-suited particularly for process synthesis problems. Furthermore, this approach can be conveniently incorporated as part of a viable, i.e. alternative strategy to solve models for decision problems involving discrete–continuous variables, i.e. with respect to more conventional or traditional techniques based on methods such as branch and bound, cutting planes, or outer approximation (Floudas and Gounaris 2008; Nagl and Marquardt 2008; Boukouvala et al. 2016).

An approach such as GDP (or a variant of such a logic-based method) is conceivably more direct and even natural in expressing the selection of conditional tasks or equipment, which is amenable to discrete decision representation. Such conditional constraints are modeled by disjunctions that employ Boolean variables to affect their existence by activating or deactivating subsets of the relevant constraints (Raman and Grossmann 1993b; Caballero and Grossmann 1999, 2011). Moreover, the logic propositions that are reformulated into linear inequalities can be used to generate strong cutting planes, which potentially offer tighter relaxations if they are suitably employed in the solution algorithm (Turkay and Grossmann 1996).

4.5.1 Propositional Logics for Superstructure Optimization Modeling

Mixed-integer programming is amenable to incorporating both qualitative and quantitative knowledge in process design and synthesis problems. Information such as that derived from engineering experience and heuristics plays an important role in design concerning decisions on process units to be included in a process network flowsheet. An example is to stipulate decisions to select treatment and conversion

Table 4.2 Common logical conditions for 0–1 variables.

Condition	Explanation
$y_1 + y_2 \leq 1$	Neither or either variable = 1 but not both
$y_1 + y_2 = 1$	Either variable values = 1 but not neither or both
$y_1 + y_2 \geq 1$	Either or both variables = 1 but not neither
$y_1 + y_2 = 2$	Both variable values = 1 but not neither or either
$y_1 - y_2 \leq 0$	Either y_2 only or both values = 1 but not y_1 alone
$y_1 - y_2 = 0$	Neither or both variables value = 1 but not either only

technologies in crude oil processing by using propositional logics formulated as algebraic constraints that use binary or zero–one (0–1) variables. These constraints are linear inequalities or equations formulated by using binary variables to represent discrete decisions for selecting alternative tasks corresponding to process units as well as alternative states corresponding to material streams.

Such logic inferences, relationships, and heuristics describing connections and interactions among process units in a superstructure can be enforced as logical constraints in mixed-integer models (MILP or MINLP) (Raman and Grossmann 1991). Logical constraints have been proven to be logic cuts that can reduce computational expense by tightening its linear relaxation and excluding fractional solutions without affecting the quality (or globality) of the optimal solution (Hooker et al. 1994). Our work employs these logical constraints to describe plausible and relevant design and structural specifications in determining an optimal refinery topology (Khor and Elkamel 2010). The underlying principles of such a logical 0–1 variable are examined and explained as a basic entity of this modeling approach.

4.5.2 Logical Binary Variables

Binary zero–one (0–1) variables enable a variety of logical conditions to be admitted in a model. Common examples of the logical conditions that can be met by combinations of two such variables (here represented by y_1 and y_2) are listed in Table 4.2.

4.5.3 Yes/No Type Binary Variables

A 0–1 variable can be used to represent yes or no decisions in an optimization model for which a value of one (1) indicates a yes decision and that of zero (0) is no. A general case is to link the 0–1 variable to a continuous decision variable in the model to represent certain states of the problem, e.g. x_i is flowrate of stream i in a process network such as that of a petroleum refinery. To allow the model solution to decide whether stream i exists in a new refinery network design, we may use a 0–1 variable to distinguish between the state in which stream i does not exist ($x_i = 0$) and that in which it exists ($x_i > 0$). For this, we include the following constraint in the model:

$$x_i \leq Uy_i, \quad y_i \in \{0,1\}, x_i \geq 0 \tag{4.1}$$

where U is a constant coefficient representing a known upper bound for the quantity of x. If an optimal solution obtained gives $y_i = 1$, then $x_i \leq U$ to admit stream i in the result (e.g. a new refinery network structure) at a level given by x_i. If $y_i = 0$, then $x_i = 0$ and the stream is excluded (from the new plant design).

Assume that x_i in a new refinery design lies between the minimum and maximum allowable quantities L and U, respectively, thus we include the following restrictions, i.e. constraints in the model:

$$\begin{aligned} x_i &\leq Uy_i \\ x_i &\geq Ly_i \\ y_i \in \{0,1\}, x_i &\geq 0 \end{aligned} \tag{4.2}$$

If an optimal solution gives $y_i = 0$, then $x_i = 0$. If $y_i = 1$, then we get the simple bounds $x_i \leq U$ and $x_i \geq L$. Similarly, if including the stream in the design is acceptable only when it can handle a certain quantity F, then the required constraint in the model is the following equality:

$$x_i = Fy_i, \quad y_i \in \{0,1\}, x_i \geq 0 \tag{4.3}$$

For this case, if an optimal solution gives $y_i = 1$, then $x_i = F$; otherwise $x_i = 0$ if $y_i = 0$ implies that the stream i does not figure in an optimal flowsheet of the proposed new refinery design.

4.5.4 Disjunctive Optimization Modeling

To obtain an equivalent mathematical representation for a propositional logic expression, we first consider basic logical operators to determine how we can transform each expression into an equivalent representation in the form of an equality (equation) or inequality. We then use these transformations to convert general logic expressions into an equivalent mathematical representation (e.g. disjunctive optimization).

The basic unit of propositional logic expression, which corresponds to a state or task, is called a literal that is a single variable that can assume either of two values: true or false. Associated with each literal P, there is another literal called NOT P, denoted as $\neg P$ such that either P or $\neg P$ is always true. A clause is a set of literals separated by OR operators and is also called a disjunction. A proposition is any logic expression that consists of a set of clauses P_i, $i \in I = 1, \ldots, N$ where N is the total number of clauses related by the logical operators OR ($\vee$), AND ($\wedge$), and IMPLICATION ($\Rightarrow$).

We assign a binary variable y_i to each proposition P_i. Its negation $\neg P_i$ denotes complement of P_i as given by $(1 - y_i)$. The logical value of true corresponds to the binary value of one and false to that of zero. We consider converting a logical proposition into its corresponding conjunctive normal form representation, which then allows us to express it as linear equality or inequality constraints. A three-step conversion procedure is as follows (Raman and Grossmann 1991):

Table 4.3 Algebraic constraints for different logical operators.

Logical operator	Logic proposition Example	Boolean expression	Algebraic constraint
OR ($\vee$)	Select at least one unit (i.e. more than one or all units)	$P_1 \vee P_2 \vee \cdots \vee P_r$	$y_1 + y_2 + \cdots + y_r \geq 1$
AND ($\wedge$)	Select all units	$P_1 \wedge P_2 \wedge \cdots \wedge P_r$	$y_1 \geq 1; y_2 \geq 1; \cdots; y_r \geq 1$
IMPLICATION ($\Rightarrow$)	Select Unit 1 only if Unit 2 is selected: P_2 only if P_1	$P_1 \Rightarrow P_2$ $\neg P_1 \vee P_2$	$(1 - y_1) + y_2 \geq 1$ $y_1 - y_2 \leq 0$ or $y_1 \leq y_2$
EQUIVALENCE ($\Leftrightarrow$)	Selecting a unit (P1) implies selecting another unit or other units, i.e. P_1 if and only if P_2: $(P_1 \Rightarrow P_2) \wedge (P_2 \Rightarrow P_1)$ or $P_1 \Leftrightarrow P_2$	$(\neg P_1 \vee P_2) \wedge (\neg P_2 \vee P_1)$	$(1 - y_1) + y_2 \geq 1$ $-y_1 + y_2 \geq 0$ $y_1 - y_2 \leq 0$ $y_1 \leq y_2$ and $(1 - y_2) + y_1 \geq 1$ $-y_2 + y_1 \geq 0$ $y_2 - y_1 \leq 0$ $y_2 \leq y_1$ or $y_1 = y_2$
EXCLUSIVE OR (EOR, $\underline{\vee}$)	Select only one unit or stream; exactly one of the variables is true	$P_1 \underline{\vee} P_2 \underline{\vee} \cdots \underline{\vee} P_r$ or P_1 EOR P_2 EOR ... EOR P_r	$y_1 + y_2 + \cdots + y_r = 1$
Classification	Select any process unit	$Q = \{P_1, P_2, \cdots, P_r\}$; Q is true if any variable is true	$y_q = y_1 + y_2 + \cdots + y_r$
Combination of EQUIVALENCE and OR	Select at least one unit (or more than one unit or all of the units) implies selecting another unit or other units	$(P_1 \vee P_2) \Leftrightarrow P_3$	$y_3 \geq y_1$ $y_3 \geq y_2$ $y_1 + y_2 \geq y_3$
Combination of EQUIVALENCE and EOR	Selection of only one process unit (or material stream) implies selecting another unit or other units	$(P_1 \underline{\vee} P_2) \Leftrightarrow P_3$	$y_1 + y_2 = y_3$

Source: Adapted from Raman and Grossmann (1991).

1) replace an implication by its equivalent disjunction:

$$P_1 \Rightarrow P_2 \Leftrightarrow \neg P_1 \vee P_2 \tag{4.4}$$

2) move the negation inward by applying DeMorgan's Theorem:

$$\begin{aligned} &\neg(P_1 \wedge P_2) \Leftrightarrow \neg P_1 \vee \neg P_2 \\ &\neg(P_1 \vee P_2) \Leftrightarrow \neg P_1 \wedge \neg P_2 \end{aligned} \tag{4.5}$$

3) recursively distribute the OR over the AND operators:

$$(P_1 \wedge P_2) \vee P_3 \Leftrightarrow (P_1 \vee P_2) \wedge (P_1 \vee P_3) \tag{4.6}$$

Converting each logical proposition into its conjunctive normal form representation, $Q_1 \wedge Q_2 \wedge \ldots \wedge Q_N$ allows expressing it as a set of linear equality and/or inequality constraints. Table 4.3 summarizes the basic operators used in logical propositions and their relations with the resulting algebraic constraints.

A generic framework of disjunctive superstructure optimization for process synthesis can be set up as consisting of the main steps shown in Figure 4.3 to obtain an optimal solution. Such a framework can be applied to a synthesis problem to derive a mathematical programming model to predict and prescribe an optimal flowsheet configuration. The initial step is to consider a suitable superstructure representation for which the three major types as introduced previously are STN (state–task network with tasks and states defined while equipment assignments are unknown explicitly), SEN (state–equipment network with tasks and equipment defined while task-to-equipment assignments are determined), and RTN (resource–task network) with combined features from the former two representation types.

Disjunctive programming forms a second step of a generic process synthesis framework to model the chosen superstructure representation as an optimization problem or a mathematical program, which may be reformulated as a MILP/MINLP. The presence of conditional tasks or equipment to be selected in a final (optimal)

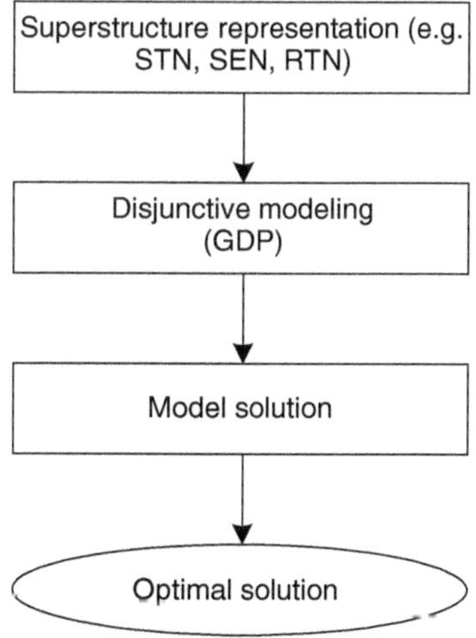

Figure 4.3 Generic disjunctive superstructure optimization framework for process synthesis.

flowsheet naturally necessitates using a discrete–continuous optimization modeling approach. Based on an associated network representation, the corresponding synthesis problem is formulated using disjunctive modeling for which suitable specific optimization formulation is available, e.g. GDP (Grossmann and Trespalacios 2013; Trespalacios and Grossmann 2014). The logic-based techniques are implemented as basis for deriving algebraic mixed-integer optimization models. Applying disjunctive modeling or programming is appropriate because process synthesis problems naturally lead to models with feasible regions of disjoint solution space while incorporating logical connectivity among the different units (tasks) and streams (states) (Raman and Grossmann 1993a).

In modeling a superstructure representation using GDP, it is necessary to identify the conditional constraints from among those that must hold for all synthesis alternatives. The conditional constraints are represented with disjunctions and assigned a Boolean variable that represents its existence (i.e. if the Boolean variable takes a value of "true"). In general, mixers and splitters are considered conditional tasks. However, if the equations that are applied to the mixers or splitters are mass and energy balances, these constraints do not involve any discrete decision (i.e. discrete variable assignment) for them to be valid.

GDP is advantageous in allowing a symbolic quantitative representation of both continuous and discrete decisions for such process synthesis problems as addressed in this book. Modeling with GDP does not explicitly involve 0–1 variables, which are more difficult to handle (conceptually and computationally) than continuous variables. In this way, it reduces the problem size by only considering disjunctions for which an associated Boolean variable is true, thus decreasing the combinatorial search effort by evaluating fewer combinations of decision variables. Moreover, GDP computations avoid the use of big-M formulations, which yield poor relaxations of nonlinearity and integrality conditions during the solution procedure (Yeomans and Grossmann 1999).

A general form of GDP model formulation is as follows (Turkay and Grossmann 1996):

$$\begin{aligned}
\min \quad & \sum_i c_i + f(x) + d^T y \\
\text{s.t.} \quad & g(x) \leq 0 \\
& Ay \geq 0 \\
& \begin{bmatrix} Y_i \\ h_i(x) \leq 0 \\ c_i = \gamma_i \end{bmatrix} \vee \begin{bmatrix} \neg Y_i \\ B^i x = 0 \\ c_i = 0 \end{bmatrix}, \quad i \in D \\
& \Omega(Y_i) = \text{true} \\
& x \in R^n, c_i \geq 0, Y_i \in \{\text{true}, \text{false}\}^m
\end{aligned} \tag{4.7}$$

The model includes disjunctions, binary variables, and integer or mixed-integer constraints. The mixed-integer nonlinear model in Eq. (4.7) involves several variable types: x is a continuous variable representing physical entities (e.g. stream flow rate, pressure, temperature, and volume), c_i is also a continuous variable representing fixed charge, and Y_i is a Boolean variable indicating unit existence by denoting whether a given disjunction i is true or false. The set of disjunctions D applies to the processing units. If a process unit exists (i.e. Y_i = true), then we enforce the equations and constraints describing that unit and apply a fixed charge, else (i.e. Y_i = false) a subset of continuous variables and fixed charge are set to zero.

Advantages of GDP for structural flowsheet optimization lie in its robustness and computational efficiency, which is comparable to algebraic MILP models and algorithms. Using disjunction in the formulation avoids the presence of big-M constraints that can lead to poor relaxation. GDP can also obviate iterations involving zero flows (Vecchietti et al. 2003).

In general, several methods are available to implement a GDP model solution including big-M reformulation as MILP; convex hull reformulation as MILP, which gives tighter relaxation than that of big-M; and using GAMS/LogMIP solver, which gives options to implement the two foregoing methods (Turkay and Grossmann 1996; Vecchietti and Grossmann 1997). More recent enhancements have allowed implementation of LogMIP with an automated mathematical program transformation (reformulation) framework called extended mathematical programming (which is available in its version 2.0 release) (GAMS Development Corporation 2019).

LogMIP which roughly stands for "logic-based mixed-integer programming" is a computer code written in C that allows specifying disjunctions in an optimization model formulation. The package can provide a general modeling framework and solution tool for handling disjunctive, algebraic, or hybrid optimization problems (Vecchietti et al. 2003).

4.6 Modeling of Process Units and Operations

4.6.1 Process Design Procedure

A process design procedure generally consists of three principal stages, namely, (i) synthesis, (ii) identification of possible need for a product, and (iii) market feasibility study. Most of the concepts and principles originate from flowsheeting activities for chemical process engineering.

4.6.2 Selecting Modeling Variables

Certain heuristics or rules of thumb can be applied in selecting appropriate modeling or design variables to build simple as well as complex models, i.e. variables should be selected with the following practical considerations (or resultant effects): choice between a number of discrete alternatives (e.g. type of heat exchanger or compressor) because it is difficult to treat them as variables to be calculated; severely constrained attributes such as maximum amount of impurity in a product, temperature in a reactor, or maximum concentration of component in an effluent stream; insensitive design to ensure saving great deal of computational requirements; save

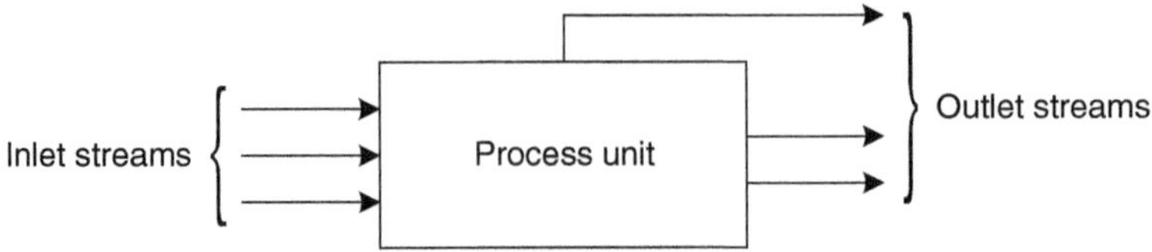

Figure 4.4 Schematic for process unit modeling.

computational effort; invariable to specifications such as pressures (or allowable pressure drops); and applies to most cases.

4.6.3 Formulating Simple Models

The key attributes modeled as optimization decision variables in formulating simple models include component molar flow rates and stream temperature and pressures. It is assumed that the outlet streams in the process units can be expressed as linear relations of the inlet streams (see Figure 4.4). Such modeling assumptions are applied, for example in the case of sharp split separators with known recovery levels or reactors with fixed conversions. The overall mass balance for process unit k is thus given by:

$$\sum_{i \in I_k} F_i = \sum_{i' \in O_k} F_{i'} \tag{4.8}$$

where F_i = flow rate of stream i in set of inlet stream I_k and stream i' in set of outlet stream O_k. The composition-based mass balance for each component j is given by the following nonlinear (bilinear) equations:

$$\sum_{i' \in I_k} \beta_{i' j,k,i} F_{i'} x_{i' j} = F_{i'} x_{i' j}, \quad \forall i \in O_k, \forall k \tag{4.9}$$

where $\beta_{i',j,k,i}$ is a constant for process unit k that gives the distribution of component j in outlet streams $i' \in O_k$ from inlet streams $i \in I_k$. For example, modeling a separator requires that $\sum_{i' \in O_k} \beta_{i' j,k,i} = 1$ and $\sum_{i' \in I_k} \beta_{i' j,k,i} = 1$. Similarly, modeling a sharp split generator requires that $|I_k| = 1$ and $|O_k| = 2$ (for the top and bottom streams) as well as $\beta_{i' j,k,i}$ is either 0 or 1 for each of the components.

4.6.4 Basic Unit Models

4.6.4.1 Mixer

A mixer is a modeling unit as shown in Figure 4.5 with no unit parameters (e.g. there is no notion of efficiency) and detailed in Table 4.4.

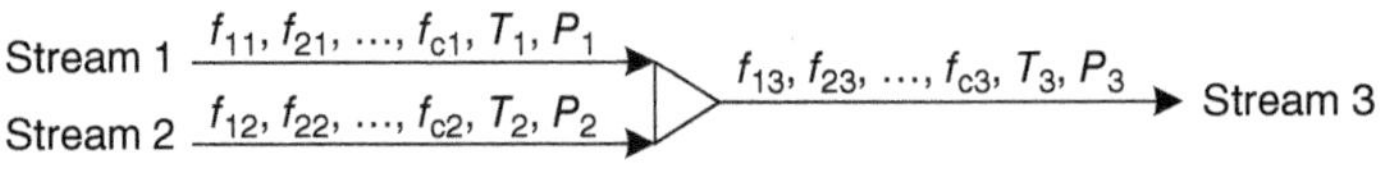

Figure 4.5 Schematic for modeling of a mixer.

Table 4.4 Degrees of freedom analysis for mixer modeling.

Equation	No. of equations	No. of variables
$f_{i3} = f_{i1} + f_{i2} \quad (i = 1, c)$	c	$3c\ (f_{ij})$
$h_3 \sum_{i=1}^{c} f_{i3} = h_1 \sum_{i=1}^{c} f_{i1} + h_2 \sum_{i=1}^{c} f_{i2}$	1	3*
$h_j = h_j(x_{ij}(i = 1, c), T_j, P_j(j = 1, 2, 3))$	3**	$3c^*\ (x_{ij}) + 6(T_j, P_j)$
$x_{ij} = f_{ij} / \sum_{i=1}^{c} f_{ij}(i = 1, c; j = 1,2, 3)$	$3c^*$	0
$P_3 = \min(P_1, P_2)$	1	0
Total equations	$4c^* + 5$	$6c + 9$
Total equations (excluding defining equations[a)])	$c + 1 + 1 = c + 2$	
Total variables (excluding intermediate variables[b)])	$3c + 6$	
Degree of freedom (No. of variables – No. of equations)	$2c + 4$	

a) Defining equations.
b) Intermediate variables.

4.6.4.2 Splitter

A key modeling attribute for representing a splitter is the split fraction ψ.

Figure 4.6 shows a schematic for modeling a stream splitter (e.g. for purge-out operation) as detailed in Table 4.5).

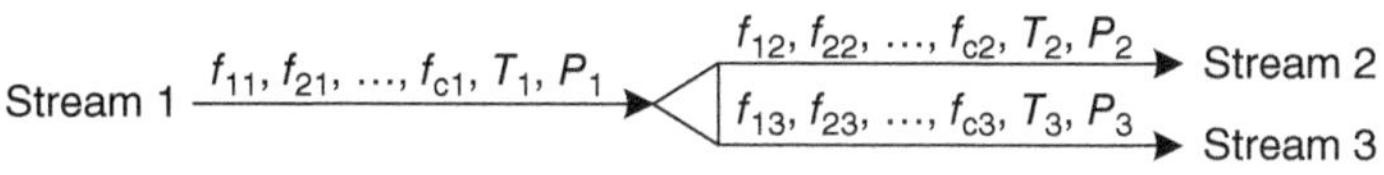

Figure 4.6 Schematic for splitter modeling.

Table 4.5 Degrees of freedom analysis for splitter modeling.

Equation	No. of equations	No. of variables
$f_{i2} = \psi f_{i1} \quad (i = 1, c)$	c	$2c{+}1\ (f_{i1}, f_{i2}, \psi)$
$f_{i3} = (1 - \psi)\psi f_{i1} \quad (i = 1, c)$	c	$c\ (f_{i3})$
$T_2 = T_3 = T_1$	2	$3\ (T_i)$
$P_2 = P_3 = P_1$	2	$3\ (P_j)$
Total no. of equations	$2c + 4$	
Total no. of variables	$3c + 7$	
Degrees of freedom	$c + 3$	

4.6.4.3 Separator

Key attributes involved in modeling a separator unit as shown in Figure 4.7 include the flash temperature and pressure detailed in Table 4.6.

4.6.4.4 Valve

A key modeling attribute for representing a valve unit is the adiabatic irreversible pressure reduction as shown in Figure 4.8 and detailed in Table 4.7, where ΔP = required pressure reduction.

4.6.4.5 Multicomponent Splitter

A multicomponent splitter model can be used as a simplified representation such as that shown in Figure 4.9 to approximate unit operations such as distillation columns

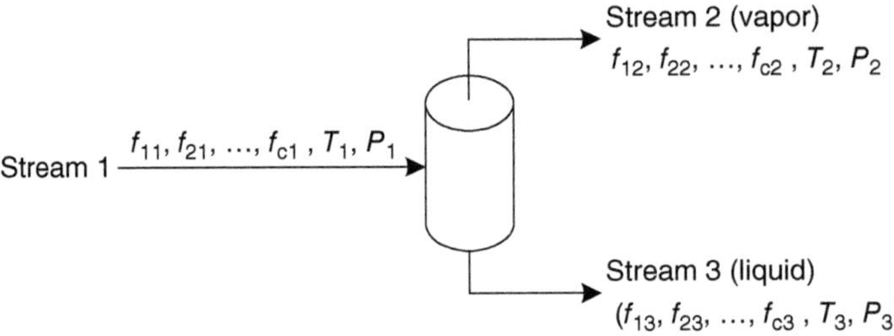

Figure 4.7 Modeling schematic for a separation vessel with a two-phase inlet stream.

Table 4.6 Degrees of freedom analysis for separators modeling.

Equation	No. of equations	No. of variables
$f_{i1} = f_{i2} + f_{i3} \quad (i = 1, c)$	c	$3c\ (f_{ij})$
$\frac{x_{i2}}{x_{i3}} = K_i \quad (i = 1, c)$	c	$3c^*\ (K_i, x_{i2}, x_{i3})$
$T_2 = T_3 = T_1$	2	$3\ (T_i)$
$P_2 = P_3 = P_1$	2	$3\ (T_i)$
$K_i = K_i(x_{i2}(i = 1, c), x_{i3}(i = 1, c), T_2, P_2)$	c**	0
$x_{ij} = \frac{f_{ij}}{\sum_{i=1}^{c} f_{ij}}, (i = 1, c; j = 2, 3)$	$2c$**	0
Total equations (excluding defining equations[a])	$2c + 4$	
Total variables (excluding intermediate variables[b])	$3c + 6$	
Degrees of freedom (No. of variables – No. of equations)	$c + 2$	

a) Defining equations.
b) Intermediate variables.

Stream 1 $f_{11}, f_{21}, \ldots, f_{c1}, T_1, P_1$ → $f_{12}, f_{22}, \ldots, f_{c2}, T_2, P_2$ → Stream 2

Figure 4.8 Modeling schematic for a valve (with adiabatic irreversible pressure reduction).

Table 4.7 Degrees of freedom analysis for modeling valves with adiabatic irreversible pressure reduction.

Equation	No. of equations	No. of variables
$f_{i2} = f_{i1} \quad (i = 1, c)$	c	$2c\,(f_{ij})$
$P_2 = P_1 - \Delta P$	1	$3\,(P_1, P_2, \Delta P)$
$h_2 = h_1$	1	$2^*\,(h_1, h_2)$
$h_j = h_j(x_{ij}(i = 1, c), T_j, P_j(j = 1, 2))$	2**	$2c^*\,(x_{ij}), 2\,(T_j, P_j)$
$x_{ij} = \frac{f_{ij}}{\sum_{i=1}^{c} f_{ij}}, (i = 1, c; j = 1, 2)$	$2c$**	0
Total equations (excluding defining equations[a)])	$c + 2$	
Total variables (excluding intermediate variables[b)])	$2c + 5$	
Degrees of freedom (No. of variables – No. of equations)	$c + 3$	

a) Defining equations.
b) Intermediate variables.

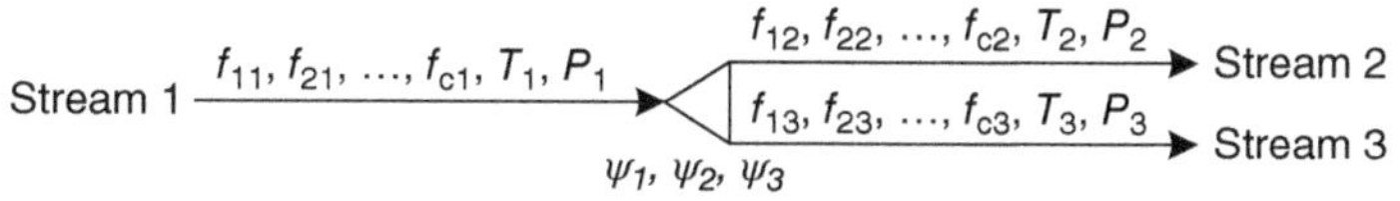

Figure 4.9 Schematic for multicomponent splitter modeling.

Table 4.8 Degrees of freedom analysis for multicomponent splitter modeling.

Equation	No. of equations	No. of variables
$f_{i2} = \psi_i f_{i1} \quad (i = 1, c)$	c	$3c\,(f_{i1}, f_{i2}, \psi_i)$
$f_{i3} = (1 - \psi_i) f_{i1} \quad (i = 1, c)$	c	$c\,(f_{i3})$
$T_2 = T_3 = T_{11}$	2	3
$P_2 = P_3 = P_1$	2	3
Total equations (excluding defining equations[a)])	$2c + 4$	
Total variables (excluding intermediate variables[b)])	$4c + 6$	
Degrees of freedom (No. of variables – No. of equations)	$2c + 2$	

a) Defining equations.
b) Intermediate variables.

(e.g. for steady-state simulation models). It might not necessarily exist as a physical unit in reality but remains useful as a modeling construct. A key modeling attribute of a multicomponent splitter is the split fraction of component i as indicated by ψ_i detailed in Table 4.8.

4.6.5 Unit Operation Models

4.6.5.1 Compressor

For the purpose of exemplifying the modeling approach presented here with more details, such a unit operation modeling technique for a compressor can be performed by considering single-stage operation under steady state and steady flow (SSF) conditions. An associated simplified representation is shown in Figure 4.10 where ΔP is the required pressure increase and η is adiabatic efficiency as described by:

$$\eta = \frac{\text{Work of reversible adiabatic compression}}{\text{Work of actual adiabatic compression}} = \frac{W^{\text{SSF}}_{\text{constant entropy}}}{W^{\text{SSF}}_{\text{actual adiabatic}}} = \frac{h_2' - h_1}{h_2 - h_1}$$

This model can be combined with a cooler model to represent a multistage compressor. A summary of the associated degrees of freedom analysis which entails in modeling such a compressor unit operation is detailed in Table 4.9.

4.6.5.2 Furnace

Key attributes of for modeling furnace unit operation include the parameters and decision variables on estimated pressure drop for process streams ΔP (which can be set to 0) and heat added to process stream Q. Schematic of a simplified representation for furnace modeling is shown in Figure 4.11. A summary of the degrees of freedom analysis for modeling this furnace unit operation is detailed in Table 4.10.

4.6.5.3 Conversion Reactor

We consider illustrating the case for a single reaction; extension to multiple independent reactions is straightforward. An example is illustrated for hydrogen bromide synthesis as described by the following equation:

$$H_2 + Br_2 \rightarrow 2HBr.$$

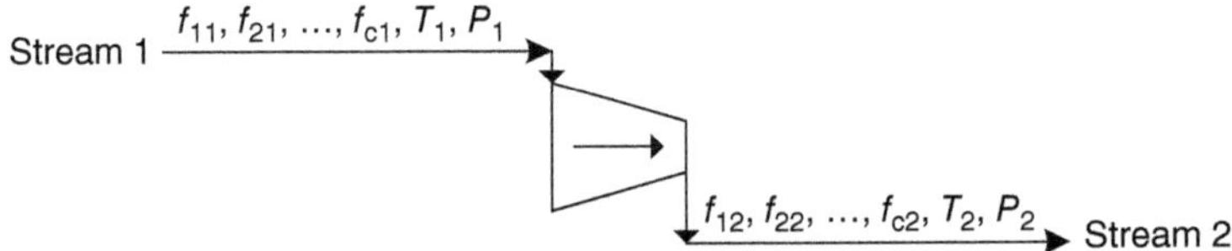

Figure 4.10 Schematic for compressor modeling.

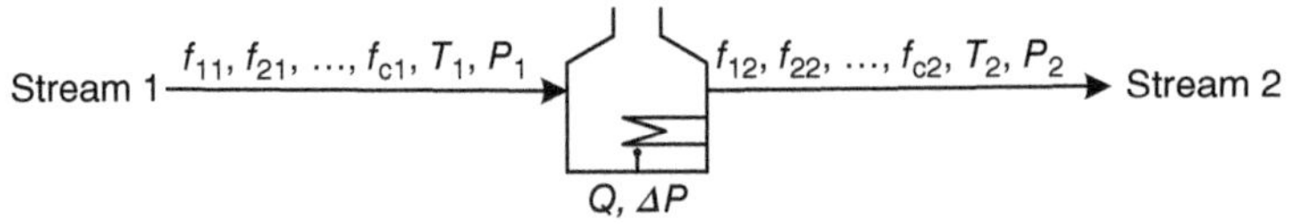

Stream 1 $f_{11}, f_{21}, \ldots, f_{c1}, T_1, P_1$ | $f_{12}, f_{22}, \ldots, f_{c2}, T_2, P_2$ Stream 2

$Q, \Delta P$

Figure 4.11 Schematic for furnace modeling.

Table 4.9 Degrees of freedom analysis for compressor modeling.

Equation	No. of equations	No. of variables
$f_{i2} = f_{i1} \quad (i = 1, c)$	c	$2c\ (f_{ij})$
$h_1 = h_1(x_{i1}(i = 1, c), T_1, P_1)$	1**	$c^*\ (x_{i1})$, $1^*\ (h_1)$, $2\ (T_1, P_1)$
$s_1 = s_1(x_{i1}(i = 1, c), T_1, P_1)$	1**	$1^*\ (s_1)$
$s'_2 = s_1$	1**	$1^*\ (s'_2)$
$P_2 = P_1 - \Delta P$	1	$2\ (P_2, \Delta P)$
$h'_2 = h'_2\ (s'_2, P_2, x_{i2}(i = 1, c))$	1**	$c^*\ (x_{i2})$
$h_2 = h_1 + \frac{h'_2 - h_1}{\eta}$	1	$1^*\ (h_2)$, $1\ (\eta)$
$h_2 = h_2(x_{i2}(i = 1, c), T_2, P_2)$	1**	$1\ (T_2)$
$x_{ij} = \frac{f_{ij}}{\sum_{i=1}^{c} f_{ij}}, (i = 1, c; j = 1, 2)$	2c**	0
$W^{SSF}_{\text{actual adiabatic}} = h_2 - h_1$	1	$1\ (W^{SSF}_{\text{actual adiabatic}})$
Total equations (excluding defining equations[a)])	$c + 3$	
Total variables (excluding intermediate variables[b)])	$2c + 7$	
Degrees of freedom (No. of variables – No. of equations)	$c + 4$	

a) Defining equations.
b) Intermediate variables.

Table 4.10 Degrees of freedom analysis for furnace modeling.

Equation	No. of equations	No. of variables
$f_{i2} = f_{i1} \quad (i = 1, c)$	c	$2c\ (f_{ij})$
$h_2 \sum_{i=1}^{c} f_{i2} = h_1 \sum_{i=1}^{c} f_{i1} + Q$	1	$2^*\ (h_1, h_2)$, $1\ (Q)$
$P_2 = P_1 - \Delta P$	1	$1\ (\Delta P)$
$h_j = h_j(x_{ij}(i = 1, c), T_j, P_j(j = 1, 2))$	2**	$4\ (T_j, P_j)$
Total equations (excluding defining equations[a)])	$c + 2$	
Total variables (excluding intermediate variables[b)])	$2c + 6$	
Degrees of freedom (No. of variables – No. of equations)	$c + 4$	

a) Defining equations.
b) Intermediate variables.

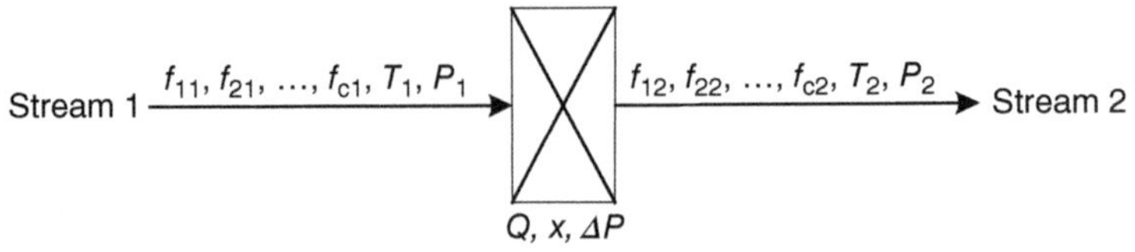

Figure 4.12 Schematic for conversion reactor modeling.

The first step is to express the reaction in the form of "Products – Reactants = 0" that can be represented as follows:

$$2\text{HBr} - \text{H}_2 - \text{Br}_2 = 0.$$

In this form, the associated stoichiometric coefficient υ_i can be defined as follows:

$$\upsilon_i = \begin{cases} > 0 & \text{for products} \\ < 0 & \text{for reactants} \\ = 0 & \text{for inerts} \end{cases}$$

The parameter called the extent of reaction ξ defines the number of moles (i.e. mole quantity) of a particular component (referred to as "key" component) with a stoichiometric coefficient equals one, which is reacting in unit time that is described by the following relations:

$$\underbrace{f_{i2}}_{\substack{\text{molar flowrate} \\ \text{leaving}}} = \underbrace{f_{i1}}_{\substack{\text{molar flowrate} \\ \text{entering}}} + \underbrace{\vartheta_i \quad \xi}_{\substack{\text{reaction} \\ \text{stoichiometric} \\ \text{coefficient}}}$$

$$f_{\text{HBr},2} = f_{\text{HBr},1} + 2\xi$$

$$f_{\text{H2},2} = f_{\text{H}_2,1} - \xi$$

$$f_{\text{Br2},2} = f_{\text{Br}_2,1} - \xi$$

$$f_{\text{inert},2} = f_{\text{inert},1}$$

Schematic of a simplified representation for conversion reactor modeling is shown in Figure 4.12 in which it is often convenient particularly when calculations are iterative to define an extent of conversion φ as a fractional conversion of the incoming amount of the key component.

A simple numerical example can be illustrated for the case if 20% of incoming H_2 is to react, then $\varphi = 0.2$ and $\xi = \varphi f_{\text{H}_2,1}$. Other parameters or variables of interest include ΔP = allowable pressure change (can be set to zero or obtain typical values from literature then design reactor to meet tolerable values), ΔH_R = heat of reaction (treated as constant, i.e. fixed-value parameter and not as variable), and Q = heat transfer rate to material flowing through reactor ($Q = 0$ for adiabatic reactor, $Q < 0$ for cooling). A summary of the degrees of freedom analysis for modeling this furnace unit operation is detailed in Table 4.11.

Table 4.11 Degrees of freedom analysis for conversion reactor modeling.

Equation	No. of equations	No. of variables
$f_{i2} = f_{i1} + \vartheta_i \xi \quad (i = 1, c)$	c	$2c + 1\ (f_{ij}, \xi)$
$h_2 \sum_{i=1}^{c} f_{i2} = h_1 \sum_{i=1}^{c} f_{i1} + (-\Delta H_R)\xi + Q$	1	$2^*\ (h_1, h_2), 1\ (Q)$
$h_j = h_j(x_{ij}(i = 1, c), T_j, P_j(j = 1, 2))$	2^{**}	$2c^*(x_{ij}), \begin{cases} 2c^*\ (x_{ij}), \\ 4(T_j, P_j) \end{cases}$
$P_2 = P_1 - \Delta P$	1	$1\ (\Delta P)$
$x_{ij} = \frac{f_{ij}}{\sum_{i=1}^{c} f_{ij}}, (i = 1, c; j = 1, 2)$	$2c^{**}$	0
Total equations (excluding defining equations[a)])	$c + 2$	
Total variables (excluding intermediate variables[b)])	$2c + 7$	
Degrees of freedom (No. of variables − No. of equations)	$(c + 2) + 3\ (Q, x, \Delta P)$	

a) Defining equations.
b) Intermediate variables.

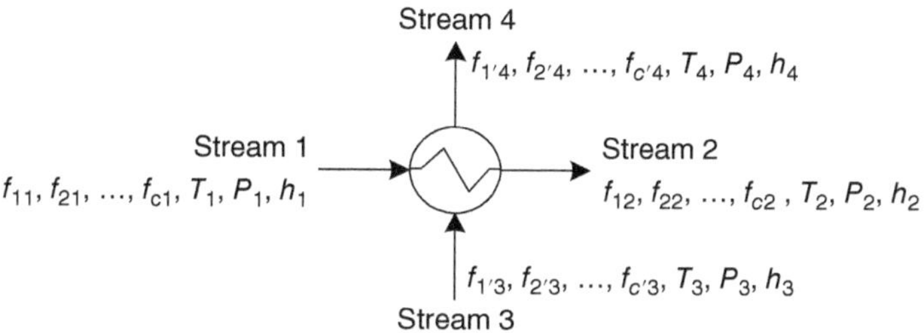

Figure 4.13 Schematic for heat exchanger modeling.

4.6.5.4 Heat Exchanger

We consider a heat exchanger model (Figure 4.13) that involves heat transfer between a stream containing c components and a stream containing c' components where ΔP and $\Delta P'$ denote allowable pressure drops for streams containing c and c' components, respectively. A summary of the degrees of freedom analysis for modeling this heat exchanger unit operation is detailed in Table 4.12.

4.6.6 Information Flow Modeling

We can perform a degree-of-freedom analysis in modeling information flow by adopting a sequential modular or an equation-oriented procedure (Grossmann and Daichendt 1996; Daichendt and Grossmann 1997). The approach involves specifying unit parameters and stream variables as exemplified through

Table 4.12 Degrees of freedom analysis for heat exchanger modeling.

Equation	No. of equations	No. of variables
$f_{i2} = f_{i1} \quad (i = 1, \ldots, c)$	c	$2c\ (f_{ij})$
$f_{i'4} = f_{i'3} \quad (i' = 1, \ldots, c')$	c'	$2c'\ (f_{ij})$
$h_2 \sum_{i=1}^{c} f_{i2} = h_1 \sum_{i=1}^{c} f_{i1} + Q$	1	$2^*\ (h_1, h_2), 1\ (Q)$
$h_4 \sum_{i'=1}^{c'} f_{i'_4} = h_3 \sum_{i'=1}^{c'} f_{i'_3} - Q$	1	$2^*\ (h_3, h_4)$
$h_j = h_j(x_{ij}(i = 1, c), T_j, P_j(j = 1, 2))$	2^{**}	$2c^*\ (x_{ij}), 4\ (T_j, P_j)$
$h'_j = h'_j\left(x_{i'j'}(i' = 1, c'), T_{j'}, P_{j'}\ (j' = 3,4)\right)$	2^{**}	$2c'^*(x_{i'j'}), 4\ (T_{j'}, P_{j'})$
$x_{ij} = \frac{f_{ij}}{\sum_{i=1}^{c} f_{ij}}, (i = 1, c; j = 1, 2)$	$2c^{**}$	0
$x_{i'j'} = \frac{f_{i'j'}}{\sum_{i'=1}^{c'} f_{i'j'}}, (i' = 1, c'; j' = 3, 4)$	$2c^{**}$	0
$P_2 = P_1 - \Delta P$	1	$1\ (\Delta P)$
$P_4 = P_3 - \Delta P'$	1	$1\ (\Delta P')$
Total equations (excluding defining equations[a)])	$c + c' + 4$	
Total variables (excluding intermediate variables[b)])	$2c + 2c' + 11$	
Degrees of freedom (No. of variables – No. of equations)	$c + c' + 7$	

a) Defining equations.
b) Intermediate variables.

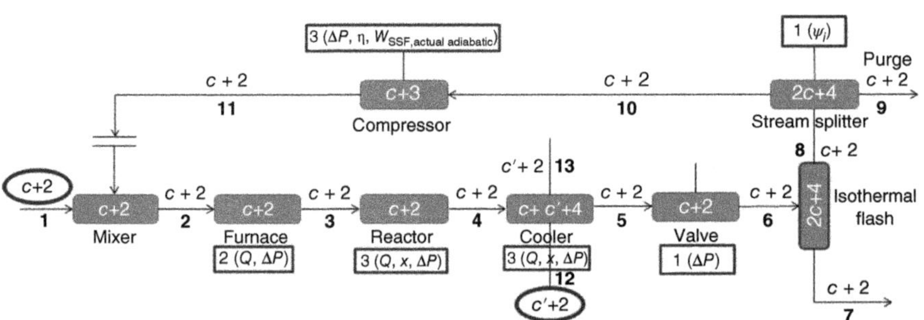

Figure 4.14 Process flowsheet example for performing degree-of-freedom analysis with specified stream variables.

flowsheeting exercise. To illustrate the approach, we consider a generic process through an example as shown in Figure 4.14. The notations used here are as follows: specified stream variables are denoted with double red concentric circles or ellipses (there is only one such variable in this example); specified unit parameters are denoted in single red circles; and directed lines correspond to the direction of information flow as shown in Figure 4.15 and detailed in Table 4.13.

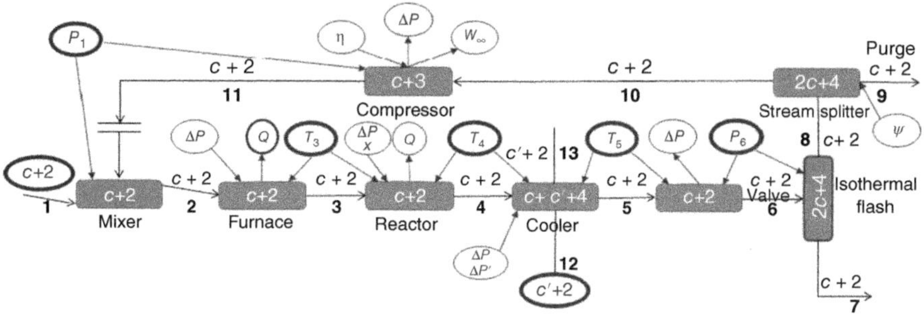

Figure 4.15 Process flowsheet example for performing degree-of-freedom analysis with specified unit parameters.

Table 4.13 Specification for perform degree-of-freedom analysis.

Specification	Attribute (variable/parameter)
Stream variable	Feed ($c + 2$), cooling stream input ($c' + 2$), reactor feed temperature (T_3), reactor exit temperature (T_4), cooler exit temperature (T_5), flash vessel pressure (P6), compressor outlet pressure (P_1) ⇒ Total number of specified stream variables $= c + c' + 9$
Unit parameter	Pressure drops over furnace, reactor, cooler (both streams); extent of conversion in reactor; purge fraction; compressor efficiency ⇒ Total number of specified unit parameters = 7
Number of independent equations associated with models	$c + 2$
Number of stream variables	$c + 2$
Total number of unit equations	$10c + c' + 23$
Total number of stream variables	$11c + 2c' + 26$
Total number of unit parameters/variables	13
Total number of variables (for units and streams)	$11c + 2c' + 26 + 13 = 11c + 2c' + 39$
Degree of freedom	$(11c + 2c' + 39) - (10c + c' + 23) = c + c' + 16$

The following example as shown in Figure 4.16 illustrates a degree-of-freedom analysis of a process flowsheet involving mass balance only and detailed in Table 4.14.

4.6.6.1 Information Flow Diagram

The following information are incorporated or reflected in an information flow diagram as exemplified in Figure 4.17:

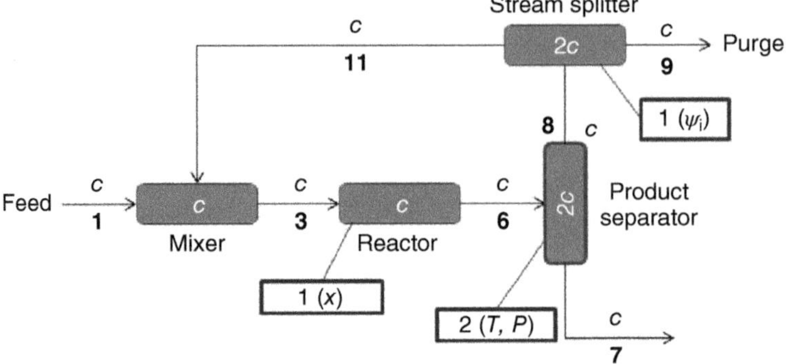

Figure 4.16 Process flowsheet example for performing degree-of-freedom analysis with mass balance only.

Table 4.14 Degrees of freedom analysis for a process flowsheet with mass balance only.

Equation	No. of equations	No. of variables
1. Mixer	c	$3c\ (f_{i1}, f_{i2}, f_{i3})$
$f_{i3} = f_{i1} + f_{i2} \quad (i = 1, \ldots, c)$		
2. Reactor	c	$c + 1\ (f_{ij}, \xi)$
$f_{i2} = f_{i1} + \xi\vartheta_i \quad (i = 1, \ldots, c)$		
3. Product separator		
$f_{i6} = f_{i7} + f_{i8} \quad (i = 1, \ldots, c)$	c	$2c\ (f_{i7}, f_{i8})$
$\frac{x_{i8}}{x_{i7}} = K_i \quad (i = 1, \ldots, c)$	c	$3c^*\ (K_i, x_{i7}, x_{i8})$
$K_i = K_i(x_{i8}(i = 1, \ldots, c), x_{i7}(i = 1, \ldots, c), T, P)$	c^{**}	$2\ (T, P)$
$x_{i8} = \frac{f_{i8}}{\sum_{i=1}^{c} f_{i8}}; x_{i7} = \frac{f_{i7}}{\sum_{i=1}^{c} f_{i7}}$	$2c^{**}$	0
4. Stream splitter		
$f_{i9} = \psi f_{i8} \quad (i = 1, c)$	c	$c + 1\ (f_{i9}, \psi)$
$f_{i11} = (1 - \psi)\psi f_{i8} \quad (i = 1, c)$	c	0
Total equations (excluding defining equations[a)])	$6c$	
Total variables (excluding intermediate variables[b)])	$7c + 4$	
Degrees of freedom (No. of variables – No. of equations)	$c + 4$	

a) Defining equations.
b) Intermediate variables.

- specified stream variables
- specified unit parameters
- number of specified design variables: c (feed molar flow rate$) + 4 unit parameters (extent of conversion, temperature and pressure in product separator, purge fraction).

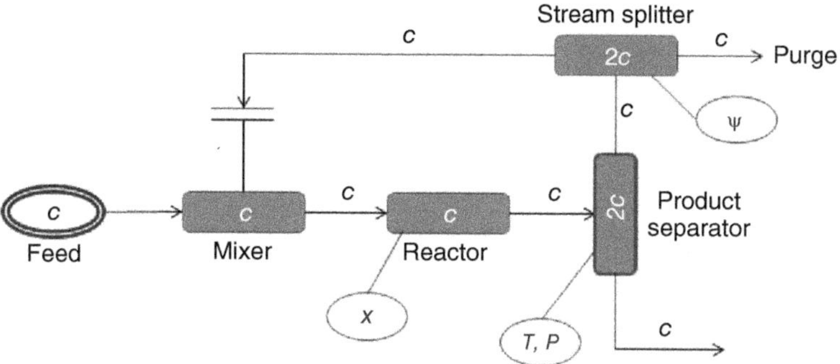

Figure 4.17 Process flowsheet example for performing degree-of-freedom analysis: information flow diagram.

4.6.6.2 Choice of Design Variables

The following heuristic rules apply to both complex and simple models:

- choose variables that involve a choice between a number of discrete alternatives (e.g. type of equipment or technology selection) – it is difficult to regard these variables to be computed;
- choose variables that are severely constrained, e.g. maximum amount of impurity in the product, maximum temperature in the reactor, and maximum concentration of a component in an effluent stream;
- choose variables (if possible) to which the design is insensitive to save much computational effort;
- choose variables that save computational effort;

It is almost invariably the practice to specify pressures and allowable pressure drops. It is also noteworthy that the first and fourth items are important and apply to most cases.

4.6.6.3 Equation Ordering

Given m variables and n equations, if $m > n$, then which variables should be selected as design variables? In which order should the equations be solved? (Must they be solved simultaneously?)

Consider the following set of equations (exemplified for a heat exchanger):

$$f_1(x_2, x_3, x_4, x_{13}) = 0$$

$$f_2(x_9, x_{10}, x_{11}, x_{12}, x_{13}) = 0$$

$$f_3(x_5, x_6) = 0$$

$$f_4(x_7, x_8) = 0$$

$$f_5(x_2, x_5, x_9, x_{10}) = 0$$

$$f_6(x_2, x_7, x_{11}, x_{12}) = 0$$

$$f_7(x_1, x_4, x_5, x_6, x_7, x_8, x_9, x_{10}x_{11}, x_{12}) = 0$$

Construct an occurrence matrix $\tilde{O}$ in which $O_{ij} = 1$ if variable j appears in equation i and is equal to 0 otherwise as represented in the following:

	x_1	x_2	x_3	x_4	x_5	x_6	x_7	x_8	x_9	x_{10}	x_{11}	x_{12}	x_{13}
f_1		1	1	1									1
f_2									1	1	1	1	1
f_3					1	1							
f_4							1	1					
f_5		1			1				1	1			
f_6		1					1				1	1	
f_7													
f_8	1			1	1	1	1	1	1	1	1	1	

The above representation consists of 7 equations and 13 variables that give 6 degrees of freedom. To specify the problem, assume that variables x_5, x_9, x_{10}, and x_{11} can be determined from the environment (e.g. fixed feed or cooling water temperature). Let us prefer x_1 to be a design variable (e.g. heat exchanger type or number of tube passes) with various levels to be considered.

Which of the remaining variables in the best choice for simplicity of computation? The first step involves deleting the variables whose values are specified, i.e. x_5, x_9, x_{10}, x_{11} , and the design variable x_1 for which various values are used to search for the best design. If a column contains only one nonzero entry, then the variable in question can only be obtained from the corresponding equation, and the value of that variable is not required to solve any other equation. Thus, the equation containing that variable may be placed last in the solution order – assign that variable as the output variable from the equation and delete the equation and the variable from $\tilde{O}$ (in the example are x_3 and f_1); now x_4 appears in f_7 only and x_{13} appears in f_2 only. The second step involves deleting x_4 and f_7, x_{13} and f_2, which leads to the following table:

	x_2	x_6	x_7	x_8	x_{12}
f_3		1			
f_4			1	1	
f_5	1				
f_6	1		1		1

x_6 appears in f_3 only
x_8 appears in f_4 only
x_{12} appears in f_6 only

The third step is to delete (x_6, f_3), (x_8, f_4), (x_{12}, f_6), which then leaves us with the following table:

	x_2	x_7
f_5	1	

It is seen that x_2 appears in f_5 only, hence the first equation to be solved is f_5 for x_2. The design variable is x_7 (i.e. note that no equation has been solved for x_4) – its value (or values) must be specified. The solution order is given as follows:

	3		
5	4	2	1
	6	7	

	[output set]
$f_5\left(x_2, x_5^*, x_9^*, x_{10}^*\right) = 0$	$\rightarrow x_2$
$f_3\left(x_5^*, x_6\right) = 0$	$\rightarrow x_6$
$f_4\left(x_7^*, x_8\right) = 0$	$\rightarrow x_8$
$f_6\left(x_2, x_7^{***}, x_{11}^*, x_{12}\right) = 0$	$\rightarrow x_{12}$
$f_2\left(x_9^*, x_{10}^*, x_{11}^*, x_{12}, x_{13}\right) = 0$	$\rightarrow x_{13}$
$f_7\left(x_1^{**}, x_4, x_5^*, x_6, x_7^{***}, x_8, x_9^*, x_{10}^*, x_{11}^*, x_{12}\right) = 0$	$\rightarrow x_4$
$f_1(x_2, x_3, x_4, x_{13}) = 0$	$\rightarrow x_3$

(where the asterisks are defined in this manner: * denotes a specified variable, ** denotes a preferred design variable, and *** denotes the design variable found by the algorithm)

An example of information flow diagram is depicted as follows:
We are given the following details:

$$f(x_1, x_2) = 0,$$

$$f(x_1, x_3) = 0,$$

$$f(x_2, x_3) = 0.$$

What are the output variables for these equations? (Try permutations.) Several observations are in place:

(a) The procedure was not dependent upon there being one degree of freedom. Even if there had been no degrees of freedom, it could still have been used to search for the order for which the equations should be solved. Had there been more than one degree of freedom, there would have been a corresponding number of undeleted variables that could have been selected as design variables.
(b) The procedure does not always terminate.

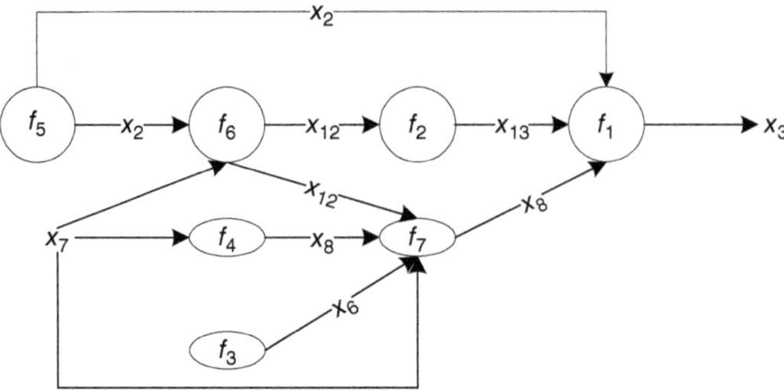

Figure 4.18 Information flow diagram: graphical depiction of the solution approach.

(c) Sometimes an equation is used to check that the value guessed for a particular variable is correct although the variable does not appear in the equation.

An instance to illustrate the aforementioned observations can be seen in the example of Figure 4.18:

	x_1	x_2	x_3	
f_1	1	1		$\tilde{O}$
f_2	1		1	
f_3		1	1	

The following trial-and-error steps are conducted:

Guess x_1:	use $f_1 \rightarrow x_2$
	use $f_2 \rightarrow x_3$
	use f_3 to check whether or not the assumed value of x_1 was correct.

If the steps do not terminate, then the following procedure is performed:

1) Determine the number of times each variable has a non-zero entry in its column, i.e. frequency $\rho(x_i)$.
2) Find the variable which has the smallest number of entries, i.e. min $\rho(x_i)$.
3) Subtract 1 from min$\rho(x_i)$, i.e. $k = \min \rho(x_i) - 1$.
4) Identify a set of k equations that have the property that when the set is selected, there remains an array containing at least one variable that appears in only one equation. If no such variable exists, increase k by one and repeat the search.
5) Delete one such set of k equations (note that k is the number of torn variables) and that the deleted equations will be the last to be solved.
6) Return to the previous algorithm.

To exemplify the aforementioned steps, suppose that in the previous example, it had been decided that x_3 would have been the design variable instead of x_9, then the following procedure applies:

	x_2	x_4	x_6	x_7	x_8	x_{12}	x_{13}
f_1	1	1					1
f_2				1		1	1
f_3			1				
f_4				1	1		
f_5	1						
f_6	1			1		1	
f_7		1	1		1	1	
$\rho(x_i)$	3	2	2	3	2	3	2

Since no variable occurs in just one equation, hence we use a new algorithm as outlined in the following:

$$\min \rho\left(x_i\right) = 2$$

$$k = \min \rho\left(x_i\right) - 1 = 1$$

Hence, it is seen that one equation has been deleted in the course of carrying out the foregoing procedure.

If f_1 is deleted, then x_4 and x_{13} appear in only one equation (note: all equations except f_5 and f_6 when deleted, lead to at least one variable appearing only once); subsequently, the following steps are carried out:

1) Delete f_1.
2) Delete x_4 and f_7, x_{13} and f_2.

Thus, this leaves the remaining of the procedure to be as follows:

	x_2	x_6	x_7	x_8	x_{12}
f_3		1			
f_4			1	1	
f_5	1				
f_6	1		1		1

3) Delete x_6 and f_3, x_8 and f_4, and x_{12} and f_6
   And this leaves the remaining steps to be as follows:

	x_2	x_7
f_5	1	

4) Delete x_2 and f_5; the undeleted variable is the torn variable x_4 solution order:

f_5	f_3	f_2	f_1
	f_4	f_7	
	f_6		

*specified variable
**torn variable

$f_5\left(x_2, x_5^*, x_9^*, x_{10}^*\right) = 0$	x_2
$f_3\left(x_5^*, x_6\right) = 0$	x_6
$f_4\left(x_7^{**}, x_8\right) = 0$	x_8 (queued variable)
$f_6\left(x_2, x_7^{***}, x_{11}^*, x_{12}\right) = 0$	x_{12}
$f_2\left(x_9^*, x_{10}^*, x_{11}^*, x_{12}, x_{13}\right) = 0$	x_{13}
$f_7\left(x_1^{**}, x_4, x_5^*, x_6, x_7^{***}, x_8, x_9^*, x_{10}^*, x_{11}^*, x_{12}\right) = 0$	x_4
$f_1(x_2, x_3, x_4, x_{13}) = 0$	CHECK VALUE of x_7 (since x_4 and x_{13} depend on the guess of x_7)

Note: Sometimes the original algorithm fails to terminate and yet there are still design variables to be chosen; it can be helpful to select a design variable at this stage rather than a torn variable

4.7 Modeling for Numerical Studies

In the past, users tend to spend substantial time on computer coding to model and solve mathematical programming problems. However, major progress has been achieved in developing mathematical optimization algorithms and computer codes that reduce significant effort and time required for model formulation and its associated implementation solution. This advancement enables focusing more time to develop robust and reliable model formulations instead of developing and coding solution procedures or customized solvers.

The formulation and solution of major mathematical programming problem types with increasingly larger scale can now be effectively and efficiently performed with modeling systems such as GAMS/General Algebraic Modeling System (Brooke et al. 1988), AIMMS (Roelofs and Bisschop 2020), and AMPL (Fourer et al. 1989). While these systems or platforms largely require expressing the model explicitly in algebraic form, they have the advantage of automatically interfacing with solver codes to handle various problem types or formulations. For example, GAMS has a library of optimization solvers (both commercial and academic) that is capable of computing globally optimum solutions for LPs, integer linear programs (ILP), and MILPs as well as determining local optima of NLPs, integer nonlinear programs (INLP), and MINLPs that have nonlinearities (convex or nonconvex) in the continuous variables (Rardin 1998). GAMS can also perform automatic differentiation and allow the use of indexed equations (constraints) that assist with helping to enhance the generation of large-scale models. Furthermore, these modeling systems are now widely available with integrated development environment (DE) on personal computers (for both desktop and laptop PC).

A platform like GAMS allows a user to focus almost exclusively on problem modeling by simplifying the model setup procedure: Define variables (i.e. decisions), equations (i.e. constraints), and data (i.e. parameters), then select a suitable solver. GAMS native setting selects a default solver that is determined to be most suited to the underlying problem structure (which chiefly entails deciding whether the model is linear or nonlinear). User can alter model formulations easily and speedily. At the outset, specifying different solvers to be tested (e.g. in exploiting the model structure) largely requires only the minimal effort of a single line coding. User can also deploy all (suitable) solution algorithms without requiring any change in the existing model, thus benefiting from the significantly reduced time that is otherwise requisite for conducting and executing computational experiments.

To reiterate, GAMS is a modeling system for optimization that provides an interface with a wide variety of different solver algorithms. Model formulations are supplied by the user to GAMS in an input file (with GAMS file extension of gms) consisting of algebraic equations formed using a high-level human-like language (in contrast with low-level machine language). GAMS then compiles the model and interfaces automatically with a solver, i.e. an optimization algorithm used entirely by itself or within a customized solution procedure. The compiled model as well as the solution or results computed by the solver is then reported back to the user through an output file (with GAMS file extension of lst). Figure 4.19 illustrates this process or workflow.

Certain GAMS model implementation features are noteworthy to highlight their advantages. The dollar control statements are lines starting with the dollar symbol ($) as compiler directives, e.g. text between $ONTEXT and $OFFTEXT are comments; $EOLCOM sets the end-of-line comment symbols (note that GAMS does not differentiate between lower- and upper-case characters in their execution). The use of dynamic sets in GAMS lends it even more flexibility amenable to developing optimization models and solution strategies (Brooke et al. 1988).

In the models presented throughout this book, the objective functions of the proposed models, whether deterministic or stochastic, are largely made up of convex functions as given in the form of a linear fixed-charge cost expression, an expectation operation, a variance operation, a recourse function, or a mean-absolute

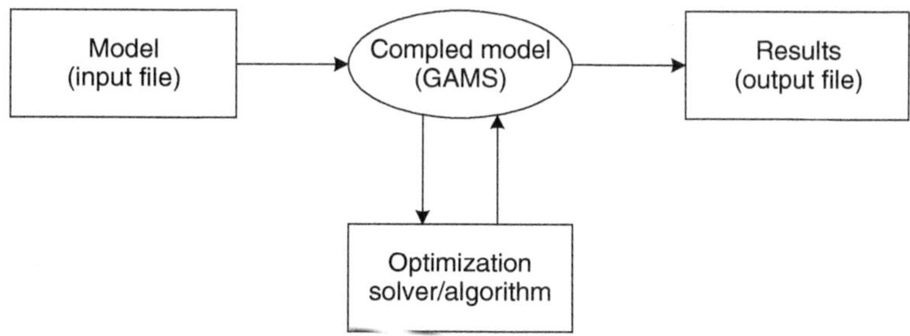

Figure 4.19 Workflow framework of GAMS modeling system.

deviation expression, mainly presented in various forms of nonnegative-weighted combinations. Therefore, we conclude that all the models possess the highly sought-after mathematical programming property of convexity based on the theorem which states that any $f(\mathbf{x})$ formed as the nonnegative-weighted ($\alpha_i \geq 0$) summation as given by $f(x) = \sum_i \alpha_i g_i(\mathbf{x})$ of convex functions $g_i(\mathbf{x})$, $i \in I$ is itself convex (Rardin 1998). Because the local optimal of a convex function is also the global optimal, a starting value to initialize the solution in GAMS is not required in our computational experiments.

4.8 Chapter Summary

This chapter presents the main modeling approach of optimization based on a superstructure (simply referred to as superstructure optimization), which is espoused and adopted throughout this book. As elaborated, this modeling framework provides ease of facilitating the modeling of appropriate logical statements based on engineering experience and design rules or heuristics to represent practical problems. We can carry out such model formulation using established concepts, techniques, and tools as well as solution methods available from the fields of mathematics, computer science, and operations research. The following chapter treats this subject (model formulation and implementation) with underlying mathematical structure in both a formal and practical manner.

References

Bagajewicz, M.J. and Manousiouthakis, V. (1992). Mass/Heat-exchanget network representation of distillation networks. *AICHE Journal* 38: 1769.

Bagajewicz, M.J., Pham, R., and Manousiouthakis, V. (1998). On the state space approach to mass/heat exchanger network design**First presented in the 1990 Annual AIChE Meeting in Chicago, paper #22d. *Chemical Engineering Science* 53 (14): 2595–2621.

Biegler, L.T., Grossmann, I.E., and Westerberg, A.W. (1997). *Systematic Methods of Chemical Process Design*. New Jersey: Prentice Hall.

Boukouvala, F., Misener, R., and Floudas, C.A. (2016). Global optimization advances in Mixed-Integer Nonlinear Programming, MINLP, and Constrained Derivative-Free Optimization, CDFO. *European Journal of Operational Research* 252 (3): 701–727.

Brooke, A., Kendrik, D., and Meeraus, A. (1988). *GAMS: A User's Guide*. Washington, DC: GAMS Development Corporation.

Caballero, J.A. and Grossmann, I.E. (1999). Aggregated models for integrated distillation systems. *Industrial & Engineering Chemistry Research* 38 (6): 2330–2344.

Caballero, J.A. and Grossmann, I.E. (2011). Logic-sequential approach to the synthesis of complex thermally coupled distillation systems. In: *21st European Symposium on Computer Aided Process Engineering*, vol. 29 (ed. E.N. Pistikopoulos, M.C. Georgiadis, and A.C. Kokossis), 211–215. Netherlands: Elsevier B.V.

Chen, Q. and Grossmann, I.E. (2017). Recent developments and challenges in optimization-based process synthesis. *Annual Review of Chemical and Biomolecular Engineering* 8 (1): 249–283.

Daichendt, M.M. and Grossmann, I.E. (1997). Integration of hierarchical decomposition and mathematical programming for the synthesis of process flowsheets. *Computers & Chemical Engineering* 22 (1-2): 147–175.

Dowling, A.W. and Biegler, L.T. (2015). A framework for efficient large scale equation-oriented flowsheet optimization. *Computers & Chemical Engineering* 72: 3–20.

Floudas, C.A., Ciric, A.R., and Grossmann, I.E. (1986). Automatic synthesis of optimum heat exchanger network configurations. *AICHE Journal* 32 (2): 276–290.

Floudas, C.A. and Gounaris, C.E. (2008). A review of recent advances in global optimization. *Journal of Global Optimization* 45 (1): 3.

Fourer, R., Gay, D.M., and Kernighan, B.W. (1989). AMPL: A Mathematical Programing Language, at Berlin, Heidelberg.

Friedler, F., Tarjan, K., Huang, Y.W., and Fan, L.T. (1993). Graph-theoretic approach to process synthesis: polynomial algorithm for maximal structure generation. *Computers & Chemical Engineering* 17 (9): 929–942.

GAMS Development Corporation. (2019). *Disjunctive Programming*. https://www.gams.com/latest/docs/UG_EMP_DisjunctiveProgramming.html (accessed 27 May 2023).

Grossmann, I.E. (2002). Review of nonlinear mixed-integer and disjunctive programming techniques. *Optimization and Engineering* 3 (3): 227–252.

Grossmann, I.E. and Daichendt, M.M. (1996). New trends in optimization-based approaches to process synthesis. *Computers & Chemical Engineering* 20 (6): 665–683.

Grossmann, I.E. and Trespalacios, F. (2013). Systematic modeling of discrete-continuous optimization models through generalized disjunctive programming. *AICHE Journal* 59 (9): 3276–3295.

Hooker, J.N., Yan, H., Grossmann, I.E., and Raman, R. (1994). Logic cuts for processing networks with fixed charges. *Computers & Operations Research* 21 (3): 265–279.

Khor, C.S. and Elkamel, A. (2010). Superstructure optimization for oil refinery design. *Petroleum Science and Technology* 28 (14): 1457–1465.

Kocis, G.R. and Grossmann, I.E. (1989). A modelling and decomposition strategy for the minlp optimization of process flowsheets. *Computers & Chemical Engineering* 13 (7): 797–819.

Kovács, Z., Ercsey, Z., Friedler, F., and Fan, L.T. (2000). Separation-network synthesis: global optimum through rigorous super-structure. *Computers & Chemical Engineering* 24 (8): 1881–1900.

Nagl, M. and Marquardt, W. (2008). *Collaborative and Distributed Chemical Engineering. From Understanding to Substantial Design Process Support*. Berlin, Heidelberg: Springer-Verlag.

Raman, R. and Grossmann, I.E. (1991). Relation between MILP modelling and logical inference for chemical process synthesis. *Computers & Chemical Engineering* 15 (2): 73–84.

Raman, R. and Grossmann, I.E. (1992). Integration of logic and heuristic knowledge in MINLP optimization for process synthesis. *Computers & Chemical Engineering* 16 (3): 155–171.

Raman, R. and Grossmann, I.E. (1993a). Relation between MILP modeling and logical inference for chemical process synthesis. *Computers & Chemical Engineering* 15: 73–84.

Raman, R. and Grossmann, I.E. (1993b). Symbolic integration of logic in mixed integer linear programming techniques for process synthesis. *Computers & Chemical Engineering* 17: 909–927.

Ramapriya, G.M., Won, W., and Maravelias, C.T. (2018). A superstructure optimization approach for process synthesis under complex reaction networks. *Chemical Engineering Research and Design* 137: 589–608.

Rardin, R.L. (1998). *Optimization in Operations Research*. New Jersey: Prentice-Hall.

Roelofs, M. and Bisschop, J. (2020). *AIMMS: The User's Guide*.

Sargent, R.W.H. and Gaminibandara, K. (1976). Optimum design of plate distillation columns. In: *Optimization in Action* (ed. L.C.W. Dixon), 267–314. London: Academic.

Trespalacios, F. and Grossmann, I.E. (2014). Review of mixed-integer nonlinear and generalized disjunctive programming methods. *Chemie Ingenieur Technik* 86 (7): 991–1012.

Trespalacios, F. and Grossmann, I.E. (2016). Cutting planes for improved global logic-based outer-approximation for the synthesis of process networks. *Computers & Chemical Engineering* 90: 201–221.

Turkay, M. and Grossmann, I.E. (1996). Logic-based MINLP algorithms for the optimal synthesis of process networks. *Computers and Chemical Engineering* 29: 959.

Vecchietti, A. and Grossmann, I.E. (1997). LOGMIP: a disjunctive 0–1 nonlinear optimizer for process systems models. *Computers & Chemical Engineering* 21: S427–S432.

Vecchietti, A., Lee, S., and Grossmann, I.E. (2003). Modeling of discrete/continuous optimization problems: characterization and formulation of disjunctions and their relaxations. *Computers & Chemical Engineering* 27 (3): 433–448.

Wu, W., Henao, C.A., and Maravelias, C.T. (2016). A superstructure representation, generation, and modeling framework for chemical process synthesis. *AICHE Journal* 62 (9): 3199–3214.

Yeomans, H. and Grossmann, I.E. (1999). A systematic modeling framework of superstructure optimization in process synthesis. *Computers & Chemical Engineering* 23: 709–731.

5

Model Formulation and Implementation

This chapter discusses a consolidation of modeling techniques and solution strategies to guide formulating optimization models amenable to be solved using readily available and reliable off-the-shelf solvers. A known technique is to supply initial values and simple bounds for the decision variables based on the informed basis such as available good data or validated reliable as well as practical solutions (Edgar et al. 2001).

5.1 Mathematical Formulation

The following presents a compact representation of a generic mixed-integer linear program (MILP)-based model formulation that is primarily employed in this work:

$$\min \quad \sum_{i,u} \mathrm{cc}_u y_{i,u} + \sum_{i,j} \mathrm{oc}_i x_{i,j} \tag{5.1}$$

$$\text{s.t.} \quad \sum_{i,j} A_{i,u} x_{i,j} = 0, \forall u \tag{5.2}$$

$$x_{i,j} \leq M_i y_{i,u}, \forall i,j \tag{5.3}$$

$$\sum_{i} B_{i,u} y_{i,u} = 0, \forall u \tag{5.4}$$

$$x_{i,j} \geq 0, \forall i,j \tag{5.5}$$

$$y_{i,u} \in \{0,1\} \tag{5.6}$$

Equation (5.1) is an objective function for cost minimization, which alternatively can be formulated as profit maximization. Equation (5.2) describes material balances on total flow rates of the stream i into unit u or flow of component j in each stream i (alternatively, it can be formulated as heat balances or design equations). Equation (5.3) describes logical constraints relating continuous variables to discrete variables by using big-M parameters (hence also called big-M constraints) based on the maximum capacity of process units as given by inlet stream flow rates. Equation (5.4) describes logical constraints stipulating design

Model-Based Optimization for Petroleum Refinery Configuration Design, First Edition. Cheng Seong Khor.

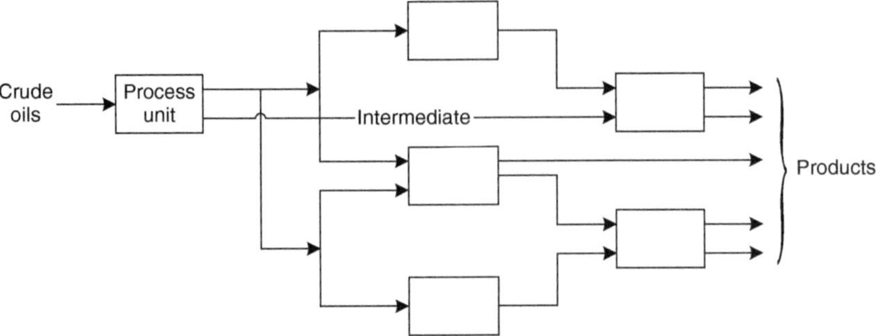

Figure 5.1 A network of processes and materials of a typical oil refinery operation.

and structural specifications for feasible selections of units and streams in process flowsheets. Equation (5.5) represents nonnegativity constraints on a vector x of continuous variables, which correspond to the state or design variables of flow rates, temperatures, pressures, compositions, and others. Equation (5.6) enforces zero–one valued restriction on a vector y of binary discrete variables to define the potential existence of a unit i or an action.

5.2 Generic Optimization Model Formulation for Refinery Planning

This section presents a basic model formulation for petroleum refinery planning. We assume that a network consisting of a set of interconnected refining process units is given in Figure 5.1 with a mixture of crude oils charge and several intermediate streams and finished products for subsequent processing (including blending) to meet specifications or for sales. The network also involves a set of materials including a mix of several crude oil types as raw material feeds besides intermediates and end products for sales.

This network can be represented by two node types, i.e. for processes and materials. The nodes are interconnected by streams to represent alternatives for material purchases, intermediates processing, and production and sales of end-products to meet market demands. Also, a finite number of time periods can be considered during which prices and demand of materials as well as capital (investment) and operating costs of the processes can vary. We devise an objective function to maximize the net present value (NPV) of the system over a specified horizon (Sahinidis et al. 1989; Escudero et al. 1999; Dempster et al. 2000; Pongsakdi et al. 2006).

We assume that the material balances in each of the process units can be expressed linearly in terms of the production rate of a main output (product from the unit) and its associated unit capacity. Also, investment costs for the processes and their expansions can be considered as linear functions of the capacities (i.e. with fixed charges).

The associated refinery planning model is subject to constraints that commonly consist of the following:

- mass balances in terms of stream quantities (volume or weight based), flow rates, and compositions of crude oils and intermediate (e.g. blending) components (typically equalities);
- capacity limits of material tanks (i.e. for crude oils, intermediates, and finished products) and process units (inequalities);
- quality specifications on finished or final products for sales based on market demand requirements, equalities, and inequalities;
- assignment of process units and tanks (equalities and inequalities involving binary variables).

5.2.1 Objective Function

A generic objective function in terms of the NPV of a refinery design project is given by:

$$\underset{S,P,E,Q}{\text{maximize}}\ \text{NPV} = \sum_{i,j,t} \left(\gamma_{i,j,t} S_{i,j,t} - \Gamma_{i,j,t} P_{i,j,t}\right) - \sum_{i,t} \left(\alpha_{i,t} E_{i,t} + \beta_{i,t} y_{i,t} Q_{i,t}\right) - \sum_{i,t} \alpha_{i,t} E_{i,t} \tag{5.7}$$

where NPV is defined as the summation of sales revenue (given by the total product sold quantity of $S_{i,j,t}$ for component j in stream i at time period t minus summation of total costs comprising purchasing cost of raw material (as given by each of the crude oils bought in quantity $P_{i,j,t}$), investment cost of new units (as given by each of the selected units with throughput $Q_{i,t}$) and expanded units (as given by each of the expanded units with additional capacity $E_{i,t}$), and several operating cost components. All the coefficients are discounted at a specified interest rate to account for the planning horizon (short-, medium-, or long-term) besides taxation effects.

5.2.2 Production Capacity and Expansion Constraints

Equation (5.8) stipulates variable lower and upper bounds for capacity expansion. A zero-value of the binary variables $y_{i,t}$ forces the capacity expansion of process i at period t to zero, i.e. $E_{i,t} = 0$; if the binary variable is equal to one, a capacity expansion between the specified bounds is executed as described in the following:

$$y_{i,t} E^L_{i,t} \leq E_{i,t} \leq y_{i,t} E^U_{i,t} \quad \forall i \in I, \forall t \in T \quad \text{where } y_{i,t} = \begin{cases} 1 & \text{if expand} \\ 0 & \text{otherwise} \end{cases} \tag{5.8}$$

Equation (5.9) defines the capacity $Q_{i,t}$ available for process i in each time period t with $Q_{i,t=0}$ representing initial capacity (which takes a zero value for non-existing processes):

$$Q_{i,t} = Q_{i,t-1} + E_{i,t}, \quad \forall i \in I, \forall t \in T \tag{5.9}$$

Constraint (5.10) expresses the condition that the operating level $W_{i,t}$ of process i in period t cannot exceed the installed capacity.

$$W_{i,t} \leq Q_{i,t}, \quad \forall i \in I, \forall t \in T \tag{5.10}$$

5.2.3 Mass Balances

Input–output mass balances around each of the process units j for component i in time period t are given by the following relations of Eqs. (5.11) and (5.12):

$$I_{i,j,t} = \mu_{i,j} W_{i,t}, \quad \forall i \in I, \forall j \in J, \forall t \in T, \tag{5.11}$$

$$O_{i,j,t} = \eta_{i,j} W_{i,t}, \quad \forall i \in I, \forall j \in J, \forall t \in T \tag{5.12}$$

where $I_{i,j,t}$ = process input of component j in stream i for time period t, and $O_{i,j,t}$ = process output of component j in stream i for time period t. Flow or flow rate (on volumetric or weight basis) of each input and output material component is proportional to the operating level of a process where $\mu_{i,j}$ and $\eta_{i,j}$ are positive fixed constants of yields, which are characteristics of each process.

Component material balance for each of the components in the entire network is described by Eq. (5.15), according to which the total amount of materials purchased from the various markets plus the amounts produced within the network (i.e. process inputs) are equal to the sum of sales and the total consumption within the network (i.e. process outputs):

$$\sum_{i \in I} P_{i,j,t} + \sum_{i \in I} O_{i,j,t} = \sum_{i \in I} \left(S_{i,j,t} + I_{i,j,t} \right), \quad \forall j \in J, \forall t \in T \tag{5.13}$$

5.2.4 Demand Constraints

Constraints in Eq. (5.14) express the lower and upper bounds on market (customer) demands for products:

$$S^{L}_{i,j,t} \leq S_{i,j,t} \leq S^{U}_{i,j,t}, \quad \forall i \in I, \forall j \in J, \forall t \in T. \tag{5.14}$$

5.2.5 Availability Constraints

Similar to Eq. (5.14), constraints (5.15) specify the lower and upper bounds on raw material availability of crude oils:

$$P^{L}_{i,j,t} \leq P_{i,j,t} \leq P^{U}_{i,j,t}, \quad \forall i \in I, \forall j \in J, \forall t \in T. \tag{5.15}$$

5.2.6 Non-Negativity Constraints

Constraints (5.16) enforce the decision variables to be non-negative:

$$E_{i,t}, Q_{i,t}, P_{i,j,t}, S_{i,j,t}, I_{i,j,t}, O_{i,j,t}, \geq 0. \tag{5.16}$$

5.3 Generic Optimization Model Formulation for Refinery Design

5.3.1 Material Balances

Material balances around the process units for a naphtha processing network can be developed in terms of total flows on the input and output streams as well as component flows incorporating the associated product yields. Balances on total mass flows around the process units are summarized in Table 5.1, while those of the component mass flows are in Table 5.2.

5.3.2 Mixed-Integer Logical Constraints

In formulating a model representation corresponding to the superstructure developed, binary variables are employed as structural variables to represent discrete decisions involved in the selection of the alternatives. These integral decisions are imposed to select process units (tasks) and material streams (states).

To ensure that the nonexistence of a process unit results in the corresponding input flow rates to the unit assuming the value of zero, we consider the formulation of

Table 5.1 Material balances in terms of total mass flow.

Unit	Material balance
ADU	$(0.4176)f_{CR} = f_{NAP1} + f_{LSRN1} + f_{HSRN1}$
HDT1	$(1.9821)(f_{HSRN2} + f_{H2\_1}) = f_{FG1} + f_{H2S1} + f_{LPG1} + f_{LSRN2} + f_{HSRN3} + f_{NAP4}$
HDT2	$(1.9821)(f_{NAP2} + f_{H2\_2}) = f_{FG2} + f_{H2S2} + f_{LPG3} + f_{LSRN3} + f_{HSRN4} + f_{NAP3}$
ISO	$f_{LSRN5} = f_{ISO} + f_{FG4}$
SRU	$f_{H2S1} + f_{H2S2} = f_S + f_{TG}$
REF	$f_{HSRN5} + f_{NAP5} = f_{H2} + f_{FG3} + f_{LPG2} + f_{REF}$
SOLD	$f_{LSRN6} + f_S + f_{GSLN} + f_{LPG5} = f_{SOLD}$
BLND	$f_{ISO} + f_{REF} = f_{GSLN}$
LPG	$f_{LPG4} = f_{LPG5}$
FGH	$f_{FG1} + f_{FG2} + f_{FG3} + f_{FG4} = f_{FG5}$
SPLT1	$f_{LSRN4} = f_{LSRN5} + f_{LSRN6}$
SPLT2	$f_{H2} = f_{H2\_1} + f_{H2\_2}$
MIX1	$f_{HSRN1} + f_{VIS1} + f_{COK1} + f_{FCC1} + f_{HCR1} + f_{PCHN1\text{-}1} = f_{HSRN2}$
MIX2	$f_{NAP1} + f_{VIS2} + f_{COK2} + f_{FCC2} + f_{HCR2} + f_{PCHN1\text{-}2} = f_{NAP2}$
MIX3	$f_{LSRN1} + f_{LSRN2} + f_{LSRN3} + f_{PCHN2} = f_{LSRN4}$
MIX4	$f_{HSRN3} + f_{HSRN4} + f_{PCHN3\text{-}1} + f_{HCR\text{-}3} = f_{HSRN5}$
MIX5	$f_{NAP3} + f_{NAP4} + f_{PCHN3\text{-}2} + f_{HCR4} = f_{NAP5}$
MIX6	$f_{LPG1} + f_{LPG2} + f_{LPG3} = f_{LPG4}$

Source: Adapted from Maples (2000).

Table 5.2 Material balances in terms of component mass flow.

Unit	Component material balance	
ADU	$(0.0555)f_{\text{CR}} = f_{\text{LSRN1}}$	$(0.2088)f_{\text{CR}} = f_{\text{NAP1}}$
	$(0.1533)f_{\text{CR}} = f_{\text{HSRN1}}$	
HDT1	$0.0109(f_{\text{H2-1}} + f_{\text{HSRN2}}) = f_{\text{FG1}}$	$0.2610(f_{\text{H2-1}} + f_{\text{HSRN2}}) = f_{\text{LSRN2}}$
	$0.0012(f_{\text{H2-1}} + f_{\text{HSRN2}}) = f_{\text{H2S1}}$	$0.7211(f_{\text{H2-1}} + f_{\text{HSRN2}}) = f_{\text{HSRN3}}$
	$0.0058(f_{\text{H2-1}} + f_{\text{HSRN2}}) = f_{\text{LPG1}}$	$0.9821(f_{\text{H2-1}} + f_{\text{HSRN2}}) = f_{\text{NAP4}}$
HDT2	$0.0109 \rightleftharpoons (f_{\text{H2-2}} + f_{\text{NAP2}}) = f_{\text{FG2}}$	$0.2610 \rightleftharpoons (f_{\text{H2-2}} + f_{\text{NAP2}}) = f_{\text{LSRN3}}$
	$0.0012 \rightleftharpoons (f_{\text{H2-2}} + f_{\text{NAP2}}) = f_{\text{H2S2}}$	$0.7211 \rightleftharpoons (f_{\text{H2-2}} + f_{\text{NAP2}}) = f_{\text{HSRN4}}$
	$0.0058 \rightleftharpoons (f_{\text{H2-2}} + f_{\text{NAP2}}) = f_{\text{LPG3}}$	$0.9821 \rightleftharpoons (f_{\text{H2-2}} + f_{\text{NAP2}}) = f_{\text{NAP3}}$
ISO	$(0.99)f_{\text{LSRN5}} = f_{\text{ISO}}$	$(0.01)f_{\text{LSRN5}} = f_{\text{FG4}}$
SRU	$(0.8478)(f_{\text{H2S1}} + f_{\text{H2S2}}) = f_{\text{S}}$	$(0.1522)(f_{\text{H2S1}} + f_{\text{H2S2}}) = f_{\text{TG}}$
REF	$0.032(f_{\text{HSRN5}} + f_{\text{NAP5}}) = f_{\text{H2}}$	$0.078(f_{\text{HSRN5}} + f_{\text{NAP5}}) = f_{\text{LPG2}}$
	$0.037(f_{\text{HSRN5}} + f_{\text{NAP5}}) = f_{\text{FG3}}$	$0.853(f_{\text{HSRN5}} + f_{\text{NAP5}}) = f_{\text{REF}}$
SPLT1	$(0.9)f_{\text{LSRN4}} = f_{\text{LSRN5}}$	$(0.1)f_{\text{LSRN4}} = f_{\text{LSRN6}}$

Source: Maples (2000), Maples (2000: p. 96), and Parkash (2003: pp. 37, 116, 225, 144).

big-*M* logical constraints to impose the relations between the continuous variables representing the flow rates of the streams and the discrete binary variables representing the existence of streams and process units. The general formulation of the big-*M* logical constraints is given by

$$f_i \leq M_i y_i \tag{5.17}$$

where f_i = flow rate (kg/day) of the output stream for process unit i, M_i = maximum capacity (kg/day) of process unit i, and y_i = existence of process unit i.

The big-*M* logical constraints are also sometimes termed as switching constraints (Rardin 1998). As mentioned, the main function of the switching constraints is to enforce the condition that no output flow exists if the associated unit does not exist. By extension, these constraints can be written as follows:

$$f_i \leq M_i z_i \tag{5.18}$$

to relate the stream flow rate to the binary variable z_i denoting the existence of the stream itself instead of the unit from where it is produced.

5.3.3 Logical Constraints on Design and Structural Specifications

Logical constraints are employed in this work to assist in determining an optimal topology of the refinery network according to an established approach (Raman and Grossmann 1991, 1992, 1993; Hooker et al. 1994) to assist in determining an optimal topology of the refinery network. The roles of the logical constraints in this regard are:

- to enforce design specifications on selecting the process units (tasks) and material streams (states) linking the units mainly based on engineering knowledge and past design experience, which describe the units' inherent characteristics;
- to enforce structural specifications stipulating interconnectivity relationships among the nodes in the network made up of units and streams, which describe the sequence in which the streams link the units.

Logical Boolean variables are employed with values of either true (equivalent to logical value of 1) or false (0) as follows: Boolean variables Y_i denote the existence of a task i in the superstructure, which involves the process units, mixers, and splitters; Boolean variables Z_i to denote the existence of a state j (sources and sinks).

5.3.4 Logic Propositional Constraints on Design Specifications

The model includes constraints formulated as logic propositions on design specifications of the refinery configuration design problem. As an example, for the following proposition to specify that ADU is selected if and only if there is crude oil feed:

$$z_{\text{CR}} = y_{\text{ADU}}, \tag{5.19}$$

the corresponding algebraic constraint representation in terms of unit existence or selection is given by

$$y_{\text{HDT1}} + y_{\text{HDT2}} - y_{\text{ADU}} = 0 \tag{5.20}$$

while the corresponding algebraic constraint representation in terms of stream existence is given by

$$z_{\text{LSRN1}} + z_{\text{NAP1}} - y_{\text{ADU}} = 0. \tag{5.21}$$

Table 5.3 shows that logical statements or logic propositions can be equivalently expressed in the form of logical expressions and clauses. The logical expressions and clauses are then transformed into algebraic integer linear inequality constraints by applying the systematic procedure proposed by Raman and Grossmann (1991). The following are two examples of transforming a logic proposition into its equivalent algebraic integer form.

5.3.4.1 Example 1

For the logical statement with "exclusive or" around atmospheric distillation unit (ADU, as given by the first item in Table 5.3):

$$Z_{\text{LSRN1}} \veebar Z_{\text{NAP1}} \Leftrightarrow Y_{\text{ADU}} \tag{5.22}$$

Since a binary variable z_i corresponds to a Boolean variable Z_i, we can rewrite the logical expression (5.22) as the following algebraic relation:

$$z_{\text{LSRN1}} + z_{\text{NAP1}} = y_{\text{ADU}}. \tag{5.23}$$

Assuming that a primary distillation unit must be present (i.e. $y_{\text{ADU}} = 1$) for a refinery to exist, the constraint given by the equation can be simplified to:

$$z_{\text{LSRN1}} + z_{\text{NAP1}} = 1. \tag{5.24}$$

Table 5.3 Logical constraints on design specifications for processing alternatives of naphtha produced from atmospheric distillation unit.

	Logical statement	Logical expression	Algebraic constraint	Desired binary variable output
1.	ADU must exist	Y_{ADU} $(Z_{LSRN1} \underline{\vee} Z_{NAP1}) \Leftrightarrow Y_{ADU}$	$y_{ADU} = 1$ $z_{LSRN1} + z_{NAP1} = y_{ADU}$	z_{LSRN1} z_{NAP1} y_{ADU} 1 0 1 0 1 1
2.	HDT1 or HDT2 exists if and only if ADU exists	$Y_{ADU} \Leftrightarrow (Y_{HDT1} \underline{\vee} Y_{HDT2})$	$y_{HDT1} + y_{HDT2} - y_{ADU} = 0$	y_{HDT1} y_{HDT2} y_{ADU} 1 0 1 0 1 1
3.	SRU exists if and only if HDT1 or HDT2 exists	$(Z_{H2S1} \underline{\vee} Z_{H2S2}) \Leftrightarrow Y_{SRU} \Leftrightarrow Y_{ADU}$	$z_{H2S1} + z_{H2S2} - y_{SRU} = 0$ $y_{SRU} = y_{ADU}$	z_{H2S1} z_{H2S2} y_{ADU} y_{SRU} 1 0 1 1 0 1 1 1
4.	MIX3 exists if and only if LSRN1 or LSRN3 exists	$(Z_{LSRN1} \vee Z_{LSRN3}) \Leftrightarrow Y_{MIX3}$	$z_{LSRN1} + z_{LSRN3} = y_{MIX3}$	z_{LSRN1} z_{LSRN3} y_{MIX3} 1 0 1 0 1 1 0 0 0

<table>
<tr><td>5.</td><td>MIX4 exists if and only if HSRN3 or HSRN4 exists</td><td>$(Z_{\text{HSRN3}} \vee Z_{\text{HSRN4}}) \Leftrightarrow Y_{\text{MIX4}}$</td><td>$z_{\text{HSRN3}} + z_{\text{HSRN4}} = y_{\text{MIX4}}$</td><td>
<table>
<tr><th>z_{HSRN3}</th><th>z_{HSRN4}</th><th>y_{MIX4}</th></tr>
<tr><td>1</td><td>0</td><td>1</td></tr>
<tr><td>0</td><td>1</td><td>1</td></tr>
<tr><td>0</td><td>0</td><td>0</td></tr>
</table>
</td></tr>
<tr><td>6.</td><td>MIX5 exists if and only if NAP3 or NAP4 exists</td><td>$(Z_{\text{NAP3}} \vee Z_{\text{NAP4}}) \Leftrightarrow Y_{\text{MIX5}}$</td><td>$z_{\text{NAP3}} + z_{\text{NAP4}} - y_{\text{MIX5}} = 0$</td><td>
<table>
<tr><th>z_{NAP3}</th><th>z_{NAP4}</th><th>y_{MIX5}</th></tr>
<tr><td>1</td><td>0</td><td>1</td></tr>
<tr><td>0</td><td>1</td><td>1</td></tr>
<tr><td>0</td><td>0</td><td>0</td></tr>
</table>
</td></tr>
<tr><td>7.</td><td>MIX3 and MIX 4, or MIX3 and MIX5, or MIX5 exist if and only if HDT1 or HDT2 exists</td><td>$(Y_{\text{HDT1}} \vee Y_{\text{HDT2}}) \Leftrightarrow ((Y_{\text{MIX3}} \wedge Y_{\text{MIX4}}) \vee ((Y_{\text{MIX3}} \wedge Y_{\text{MIX5}}) \vee Y_{\text{MIX5}}$</td><td>$2y_{\text{MIX3}} + y_{\text{MIX5}} \geq y_{\text{HDT1}}$
$y_{\text{MIX3}} + y_{\text{MIX4}} + y_{\text{MIX5}} \geq y_{\text{HDT1}}$
$y_{\text{MIX3}} + 2y_{\text{MIX5}} \geq y_{\text{HDT1}}$
$y_{\text{MIX3}} + 2y_{\text{MIX5}} \geq y_{\text{HDT1}}$
$y_{\text{MIX4}} + 2y_{\text{MIX5}} \geq y_{\text{HDT1}}$

$2y_{\text{MIX3}} + y_{\text{MIX5}} \geq y_{\text{HDT2}}$
$y_{\text{MIX3}} + y_{\text{MIX4}} + y_{\text{MIX5}} \geq y_{\text{HDT2}}$
$y_{\text{MIX3}} + 2y_{\text{MIX5}} \geq y_{\text{HDT2}}$
$y_{\text{MIX3}} + 2y_{\text{MIX5}} \geq y_{\text{HDT2}}$
$y_{\text{MIX4}} + 2y_{\text{MIX5}} \geq y_{\text{HDT2}}$</td><td>
<table>
<tr><th>y_{HDT1}</th><th>y_{HDT2}</th><th>y_{MIX3}</th><th>y_{MIX4}</th><th>y_{MIX5}</th></tr>
<tr><td>1</td><td>0</td><td>1</td><td>1</td><td>0</td></tr>
<tr><td>1</td><td>0</td><td>1</td><td>0</td><td>1</td></tr>
<tr><td>0</td><td>1</td><td>1</td><td>1</td><td>0</td></tr>
<tr><td>0</td><td>1</td><td>0</td><td>0</td><td>1</td></tr>
<tr><td>0</td><td>0</td><td>0</td><td>0</td><td>0</td></tr>
</table>
</td></tr>
</table>

Table 5.3 (Continued)

<table>
<tr><th colspan="2">Logical statement</th><th>Logical expression</th><th>Algebraic constraint</th><th>Desired binary variable output</th></tr>
<tr><td>8.</td><td>REF exists if and only if HSRN5 or NAP5 exists</td><td>$(Z_{HSRN5} \veebar Z_{NAP5}) \Leftrightarrow Y_{REF} \Leftrightarrow Y_{ADU}$</td><td>$y_{REF} = y_{ADU}$
$z_{HSRN5} + z_{NAP5} - y_{REF} = 0$</td><td><table>
<tr><th>z_{HSRN5}</th><th>z_{NAP5}</th><th>y_{REF}</th><th>y_{ADU}</th></tr>
<tr><td>1</td><td>0</td><td>1</td><td>1</td></tr>
<tr><td>0</td><td>1</td><td>1</td><td>1</td></tr>
</table></td></tr>
<tr><td>9.</td><td>LPG exists if and only if HDT1 or HDT2 exists</td><td>$(Y_{HDT1} \vee Y_{HDT2}) \Leftrightarrow Y_{LPG} \Leftrightarrow Y_{ADU}$
$(Z_{LPG1} \vee Z_{LPG2} \vee Z_{LPG3}) \Leftrightarrow Y_{LPG}$</td><td>$y_{LPG} = y_{ADU} = 1$
$y_{HDT1} + y_{HDT2} - y_{LPG} = 0$
$y_{HDT1} + y_{HDT2} = 1$
$z_{LPG1} + z_{LPG2} + z_{LPG3} - y_{LPG} = 1$
$y_{HDT1} + y_{HDT2} = z_{LPG1} + z_{LPG2} + z_{LPG3} - y_{LPG}$</td><td><table>
<tr><th>z_{LPG1}</th><th>z_{LPG2}</th><th>z_{LPG3}</th><th>y_{LPG}</th><th>y_{HDT1}</th><th>y_{HDT2}</th></tr>
<tr><td>1</td><td>1</td><td>0</td><td>1</td><td>1</td><td>0</td></tr>
<tr><td>0</td><td>1</td><td>1</td><td>1</td><td>0</td><td>1</td></tr>
</table></td></tr>
<tr><td>10.</td><td>FGH exists if and only if HDT1, HDT2, or ISO exist</td><td>$(Z_{FG1} \vee Z_{FG2} \vee Z_{FG3} \vee Z_{FG4}) \Leftrightarrow Y_{FGH}$</td><td>$y_{FGH} \geq z_{FG1}$
$y_{FGH} \geq z_{FG2}$
$y_{FGH} \geq z_{FG3}$
$y_{FGH} \geq z_{FG4}$
$z_{FG1} + z_{FG2} + z_{FG3} + z_{FG4} \geq y_{FGH}$</td><td><table>
<tr><th>z_{FG1}</th><th>z_{FG2}</th><th>z_{FG3}</th><th>z_{FG4}</th><th>y_{FGH}</th></tr>
<tr><td>1</td><td>0</td><td>1</td><td>0</td><td>1</td></tr>
<tr><td>1</td><td>0</td><td>1</td><td>1</td><td>1</td></tr>
<tr><td>0</td><td>0</td><td>1</td><td>1</td><td>1</td></tr>
<tr><td>0</td><td>1</td><td>1</td><td>0</td><td>1</td></tr>
<tr><td>0</td><td>1</td><td>1</td><td>1</td><td>1</td></tr>
</table></td></tr>
</table>

11.	ISO exists if and only if HDT1 exists	$Y_{\text{ISO}} \Leftrightarrow Y_{\text{HDT1}}$	$y_{\text{ISO}} = y_{\text{HDT1}}$	y_{ISO} | y_{HDT1} 1 | 1 0 | 0
12.	BLND exists if and only if Z_{ISO} and/or Z_{REF} exist	$(Z_{\text{ISO}} \vee Z_{\text{REF}}) \Leftrightarrow Y_{\text{BLND}}$	$y_{\text{BLND}} \geq z_{\text{ISO}}$ $y_{\text{BLND}} \geq z_{\text{REF}}$ $z_{\text{ISO}} + z_{\text{REF}} \geq y_{\text{BLND}}$	z_{ISO} | z_{REF} | y_{BLND} 1 | 1 | 1 1 | 0 | 0 0 | 1 | 0
13.	SOLD exists if and only if ADU exists	$Y_{\text{SOLD}} = Y_{\text{ADU}}$	$y_{\text{SOLD}} = y_{\text{ADU}}$	y_{SOLD} | y_{ADU} 1 | 1 0 | 0

5.3.4.2 Example 2

For the logical statement with "or" for fuel gas header (FGH as given by an item [listed as the tenth] in Table 5.3), we are given the following condition:

$$\begin{aligned}
&\left(Z_{\text{FG1}\rightleftharpoons} \vee Z_{\text{FG2}\rightleftharpoons} \vee Z_{\text{FG3}\rightleftharpoons} \vee Z_{\text{FG4}\rightleftharpoons}\right) \Leftrightarrow Y_{\text{FGH}\rightleftharpoons}\\
&\left\{\left(Z_{\text{FG}\rightleftharpoons 1\rightleftharpoons} \vee Z_{\text{FG2}\rightleftharpoons} \vee Z_{\text{FG3}\rightleftharpoons} \vee Z_{\text{FG4}\rightleftharpoons}\right) \Rightarrow Y_{\text{FGH}\rightleftharpoons}\right\} \wedge \left\{Y_{\text{FGH}\rightleftharpoons}\right.\\
&\left.\Rightarrow \left(Z_{\text{FG1}\rightleftharpoons} \vee Z_{\text{FG2}\rightleftharpoons} \vee Z_{\text{FG3}\rightleftharpoons} \vee Z_{\text{FG4}\rightleftharpoons}\right)\right\}.
\end{aligned} \tag{5.25}$$

First, consider the following proposition in terms of Boolean variables:

$$\begin{aligned}
&\left(Z_{\text{FG1}\rightleftharpoons} \vee Z_{\text{FG2}\rightleftharpoons} \vee Z_{\text{FG3}\rightleftharpoons} \vee Z_{\text{FG4}\rightleftharpoons}\right) \Rightarrow Y_{\text{FGH}\rightleftharpoons}\\
&\neg\left(Z_{\text{FG1}\rightleftharpoons} \vee Z_{\text{FG2}\rightleftharpoons} \vee Z_{\text{FG3}\rightleftharpoons} \vee Z_{\text{FG4}\rightleftharpoons}\right) \vee Y_{\text{FGH}\rightleftharpoons}\\
&\left(\neg Z_{\text{FG1}\rightleftharpoons} \wedge \neg Z_{\text{FG2}\rightleftharpoons} \wedge \neg Z_{\text{FG3}\rightleftharpoons} \wedge \neg Z_{\text{FG4}\rightleftharpoons}\right) \vee Y_{\text{FGH}\rightleftharpoons}
\end{aligned} \tag{5.26}$$

with the corresponding algebraic constraints denoted as follows:

$$\begin{aligned}
&\underbrace{\left(\neg Z_{\text{FG1}\rightleftharpoons} \vee Y_{\text{FGH}\rightleftharpoons}\right)}_{\substack{1 - z_{\text{FG1}\rightleftharpoons} + y_{\text{FGH}\rightleftharpoons} \geq 1 \\ y_{\text{FGH}\rightleftharpoons} \geq z_{\text{FG1}\rightleftharpoons}}} \wedge \underbrace{\left(\neg Z_{\text{FG2}\rightleftharpoons} \vee Y_{\text{FGH}\rightleftharpoons}\right)}_{y_{\text{FGH}\rightleftharpoons} \geq z_{\text{FG2}}}\\
&\wedge \underbrace{\left(\neg Z_{\text{FG3}\rightleftharpoons} \vee Y_{\text{FGH}\rightleftharpoons}\right)}_{y_{\text{FGH}\rightleftharpoons} \geq z_{\text{FG3}\rightleftharpoons}} \wedge \underbrace{\left(\neg Z_{\text{FG4}\rightleftharpoons} \vee Y_{\text{FGH}\rightleftharpoons}\right)}_{y_{\text{FGH}\rightleftharpoons} \geq z_{\text{FG4}\rightleftharpoons}}.
\end{aligned} \tag{5.27}$$

Thus, reformulation gives the following algebraic constraints:

$$\begin{aligned}
&y_{\text{FGH}\rightleftharpoons} \geq z_{\text{FG1}\rightleftharpoons}\\
&y_{\text{FGH}\rightleftharpoons} \geq z_{\text{FG2}}\\
&y_{\text{FGH}\rightleftharpoons} \geq z_{\text{FG3}\rightleftharpoons}\\
&y_{\text{FGH}\rightleftharpoons} \geq z_{\text{FG4}\rightleftharpoons}.
\end{aligned} \tag{5.28}$$

Then consider the proposition as follows:

$$\begin{aligned}
Y_{\text{FGH}\rightleftharpoons} \Rightarrow &\left(Z_{\text{FG1}\rightleftharpoons} \vee Z_{\text{FG2}\rightleftharpoons} \vee Z_{\text{FG3}\rightleftharpoons} \vee Z_{\text{FG4}\rightleftharpoons}\right)\\
&\neg Y_{\text{FGH}\rightleftharpoons} \vee \left(Z_{\text{FG1}\rightleftharpoons} \vee Z_{\text{FG2}\rightleftharpoons} \vee Z_{\text{FG3}\rightleftharpoons} \vee Z_{\text{FG4}\rightleftharpoons}\right)\\
&\left(1 - y_{\text{FGH}}\right) + z_{\text{FG1}} + z_{\text{FG2}} + z_{\text{FG3}} + z_{\text{FG4}} \geq 1\\
&z_{\text{FG1}} + z_{\text{FG2}} + z_{\text{FG3}} + z_{\text{FG4}} \geq y_{\text{FGH}},
\end{aligned} \tag{5.29}$$

therefore, the resulting propositions are as given in the following:

$$\begin{aligned}
Y_{\text{HDT}\rightleftharpoons 1\rightleftharpoons} &\rightleftharpoons \Rightarrow Z_{\text{HSRN2}\rightleftharpoons} \vee Z_{\text{H2-1}}\\
Z_{\text{HSRN}\rightleftharpoons 2\rightleftharpoons} &\Rightarrow Y_{\text{HDT}\rightleftharpoons 1\rightleftharpoons}\\
Z_{\text{H2-1}} &\Rightarrow Y_{\text{HDT1}\rightleftharpoons}\\
Y_{\text{HDT}\rightleftharpoons 1\rightleftharpoons} &\Rightarrow Z_{\text{H2S}\rightleftharpoons 1} \vee Z_{\text{LPG}\rightleftharpoons 1} \vee Z_{\text{LSRN}\rightleftharpoons 2} \vee Z_{\text{HSRN}\rightleftharpoons 3\rightleftharpoons} \vee Z_{\text{NAP}\rightleftharpoons 4} \vee Z_{\text{FG}\rightleftharpoons 1}\\
Z_{\text{H2S}\rightleftharpoons 1} &\Rightarrow Y_{\text{HDT}\rightleftharpoons 1\rightleftharpoons}\\
Z_{\text{LPG}\rightleftharpoons 1} &\Rightarrow Y_{\text{HDT}\rightleftharpoons 1\rightleftharpoons}\\
Z_{\text{LSRN}\rightleftharpoons 2} &\Rightarrow Y_{\text{HDT}\rightleftharpoons 1\rightleftharpoons}
\end{aligned}$$

$$
\begin{aligned}
Z_{\text{HSRN}\rightleftharpoons 3\rightleftharpoons} &\Rightarrow Y_{\text{HDT}\rightleftharpoons 1\rightleftharpoons}\\
Z_{\text{NAP}\rightleftharpoons 4} &\Rightarrow Y_{\text{HDT}\rightleftharpoons 1\rightleftharpoons}\\
Z_{\text{FG}\rightleftharpoons 1} &\Rightarrow Y_{\text{HDT}\rightleftharpoons 1\rightleftharpoons}
\end{aligned}
\tag{5.30}
$$

5.3.5 Logic Propositional Constraints on Structural Specifications

The structural specifications enforce the interconnectivity among the states and the tasks in the superstructure. The logical constraints on structural specifications employed and their transformation to algebraic integer linear inequality constraints are listed in Table 5.4.

5.3.6 Generalized Disjunctive Programming

Generalized disjunctive programming (GDP) formulation is given by:

$$
\begin{aligned}
\min\ & Z = \sum_i c_i + f(x) + d^T y\\
\text{s.t. } & g(x) \leq 0\\
& r(x) + D(y) \leq 0\\
& Ay \geq 0\\
& \begin{bmatrix} Y_{ik} \\ h_i(x) \leq 0 \\ c_i = \gamma_i \end{bmatrix} \vee \begin{bmatrix} \neg Y_{ik} \\ B^i x = 0 \\ c_i = 0 \end{bmatrix}, i \in D, k \in K\\
& \Omega\left(Y_{ik}\right) = \text{True}\\
& x \in \Re^n, c_i \geq 0, Y_i \in \{\text{True, False}\}^m
\end{aligned}
\tag{5.31}
$$

where x is continuous variables on flows and other process attributes, y is binary variables on unit selection, Y_{it} is Boolean variables associated with unit existences for a set of disjunctions $k \in K$, $\Omega(Y_{ik})$ is propositional logics involving only Boolean variables, c_i are continuous variables of costs for each of the disjunctions k that are activated to fixed charges of values c_i if the corresponding term of the disjunction is true, $g(x)$ are linear or nonlinear inequalities independent of the discrete choices, $f(x)$ is a linear or nonlinear objective function, $r(x) + D(y) \leq 0$ are general mixed-integer algebraic equations, $Ay \geq 0$ is a set of integer inequalities, and $d^T y$ is linear cost terms.

We apply two approaches to formulate the conditional constraints in a GDP model. Approach 1 involves direct formulation on the process unit existences as exemplified in the following for atmospheric distillation unit (ADU):

$$
\begin{bmatrix} Y_{\text{ADU}} \\ (0.2088) \rightleftharpoons f_{\text{CR}} = f_{\text{NAP1}} \\ (0.0555) \rightleftharpoons f_{\text{CR}} = f_{\text{LSRN1}} \\ (0.1533) \rightleftharpoons f_{\text{CR}} = f_{\text{HSRN1}} \\ f_{\text{NAP1}} = f_{\text{LSRN1}} + f_{\text{HSRN1}} \\ c_{\text{ADU}} = 228 \end{bmatrix} \vee \begin{bmatrix} \neg Y_{\text{ADU}} \\ f_{\text{CR}} = 0 \\ f_{\text{NAP1}} = 0 \\ f_{\text{LSRN1}} = 0 \\ f_{\text{HSRN1}} = 0 \\ c_{\text{ADU}} = 0 \end{bmatrix}
\tag{5.32}
$$

Table 5.4 Logic propositions and their equivalent algebraic integer constraints for structural specifications.

Unit	Logic proposition	Algebraic constraint
ADU	$\left.\begin{array}{l} Z_{CR} \Rightarrow Y_{ADU} \\ Y_{ADU} \Rightarrow Z_{CR} \end{array}\right\} Z_{CR} \Leftrightarrow Y_{ADU}$	$\neg Z_{CR} \vee Y_{ADU}$
		$1 - z_{CR} + y_{ADU} \geq 1$
		$z_{CR} \leq y_{ADU}$
		and
		$\neg Y_{ADU} \vee Z_{CR}$
		$1 - y_{ADU} + z_{CR} \geq 1$
		$y_{ADU} \leq z_{CR}$
	$Y_{ADU} \Rightarrow Z_{LSRN1} \vee Z_{HSRN1} \vee Z_{NAP1}$	$y_{ADU} \leq z_{LSRN1} + z_{HSRN1} + z_{NAP1}$
	$Z_{LSRN1} \Rightarrow Y_{ADU}$	$z_{LSRN1} \leq y_{ADU}$
	$Z_{HSRN1} \Rightarrow Y_{ADU}$	$z_{HSRN1} \leq y_{ADU}$
	$Z_{NAP1} \Rightarrow Y_{ADU}$	$z_{NAP1} \leq y_{ADU}$
HDT1	$Y_{HDT1} \Rightarrow Z_{HSRN2} \vee Z_{H2\text{-}1}$	$y_{HDT1} \leq z_{HSRN2} + z_{H2\text{-}1}$
	$Z_{HSRN2} \Rightarrow Y_{HDT1}$	$z_{HSRN2} \leq y_{HDT1}$
	$Z_{H2\text{-}1} \Rightarrow Y_{HDT1}$	$z_{H2\text{-}1} \leq y_{HDT1}$
	$Y_{HDT1} \Rightarrow Z_{H2S1} \vee Z_{LPG1} \vee Z_{LSRN2} \vee Z_{HSRN3} \vee Z_{NAP4} \vee Z_{FG1}$	$y_{HDT1} \leq z_{H2S1} + z_{LPG1} + z_{LSRN2} + z_{HSRN3} + z_{NAP4} + z_{FG1}$
	$Z_{H2S1} \Rightarrow Y_{HDT1}$	$z_{H2S1} \leq y_{HDT1}$
	$Z_{LPG1} \Rightarrow Y_{HDT1}$	$z_{LPG1} \leq y_{HDT1}$
	$Z_{LSRN2} \Rightarrow Y_{HDT1}$	$z_{LSRN2} \leq y_{HDT1}$
	$Z_{HSRN3} \Rightarrow Y_{HDT1}$	$z_{HSRN3} \leq y_{HDT1}$
	$Z_{NAP4} \Rightarrow Y_{HDT1}$	$z_{NAP4} \leq y_{HDT1}$
	$Z_{FG1} \Rightarrow Y_{HDT1}$	$z_{FG1} \leq y_{HDT1}$
HDT2	$Y_{HDT2} \Rightarrow Z_{NAP2} \vee Z_{H2\text{-}2}$	$y_{HDT2} \leq z_{NAP2} + z_{H2\text{-}2}$
	$Z_{NAP2} \Rightarrow Y_{HDT2}$	$z_{NAP2} \leq y_{HDT2}$
	$Z_{H2\text{-}2} \Rightarrow Y_{HDT2}$	$z_{H2\text{-}2} \leq y_{HDT2}$
	$Y_{HDT2} \Rightarrow Z_{H2S2} \vee Z_{LPG3} \vee Z_{LSRN3} \vee Z_{HSRN4} \vee Z_{NAP3} \vee Z_{FG2}$	$y_{HDT2} \leq z_{H2S2} + z_{LPG3} + z_{LSRN3} + z_{HSRN4} + z_{NAP3} + z_{FG2}$
	$Z_{H2S2} \Rightarrow Y_{HDT2}$	$z_{H2S2} \leq y_{HDT2}$
	$Z_{LPG3} \Rightarrow Y_{HDT2}$	$z_{LPG3} \leq y_{HDT2}$
	$Z_{LSRN3} \Rightarrow Y_{HDT2}$	$z_{LSRN3} \leq y_{HDT2}$
	$Z_{HSRN4} \Rightarrow Y_{HDT2}$	$z_{HSRN4} \leq y_{HDT2}$
	$Z_{NAP3} \Rightarrow Y_{HDT2}$	$z_{NAP3} \leq y_{HDT2}$
	$Z_{FG2} \Rightarrow Y_{HDT2}$	$z_{FG2} \leq y_{HDT2}$
ISO	$Z_{LSRN5} \Rightarrow Y_{ISO}$	$z_{LSRN5} \leq y_{ISO}$
	$Y_{ISO} \Rightarrow Z_{LSRN5}$	$y_{ISO} \leq z_{LSRN5}$
	$Y_{ISO} \Rightarrow Z_{ISO} \vee Z_{FG4}$	$y_{ISO} \leq z_{ISO} + z_{FG4}$
	$Z_{ISO} \Rightarrow Y_{ISO}$	$z_{ISO} \leq y_{ISO}$
	$Z_{FG4} \Rightarrow Y_{ISO}$	$z_{FG4} \leq y_{ISO}$

(continued)

Table 5.4 (Continued)

Unit	Logic proposition	Algebraic constraint
SRU	$Y_{SRU} \Rightarrow Z_{H2S1} \vee Z_{H2S2}$	$y_{SRU} \leq z_{H2S1} + z_{H2S2}$
	$Z_{H2S1} \Rightarrow Y_{SRU}$	$z_{H2S1} \leq y_{SRU}$
	$Z_{H2S2} \Rightarrow Y_{SRU}$	$z_{H2S2} \leq y_{SRU}$
	$Y_{SRU} \Rightarrow Z_{S} \vee Z_{TG}$	$y_{SRU} \Rightarrow z_{S} + z_{TG}$
	$Z_{S} \Rightarrow Y_{SRU}$	$z_{S} \leq y_{SRU}$
	$Z_{TG} \Rightarrow Y_{SRU}$	$z_{TG} \leq y_{SRU}$
REF	$Y_{REF} \Rightarrow Z_{HSRN5} \vee Z_{NAP5}$	$y_{REF} \leq z_{HSRN5} + z_{NAP5}$
	$Z_{HSRN5} \Rightarrow Y_{REF}$	$z_{HSRN5} \leq y_{REF}$
	$Z_{NAP5} \Rightarrow Y_{REF}$	$z_{NAP5} \leq y_{REF}$
	$Y_{REF} \Rightarrow Z_{H2} \vee Z_{FG3} \vee Z_{LPG2} \vee Z_{REF}$	$y_{REF} \leq z_{H2} + z_{FG3} + z_{LPG2} + z_{REF}$
	$Z_{H2} \Rightarrow Y_{REF}$	$z_{H2} \leq y_{REF}$
	$Z_{FG3} \Rightarrow Y_{REF}$	$z_{FG3} \leq y_{REF}$
	$Z_{LPG2} \Rightarrow Y_{REF}$	$z_{LPG2} \leq y_{REF}$
	$Z_{REF} \Rightarrow Y_{REF}$	$z_{REF} \leq y_{REF}$
SOLD	$Y_{SOLD} \Rightarrow Z_{LSRN6} \vee Z_{GSLN} \vee Z_{S} \vee Z_{LPG5}$	$y_{SOLD} \leq z_{LSRN6} + z_{GSLN} + z_{S} + z_{LPG5}$
	$Z_{LSRN6} \Rightarrow Y_{SOLD}$	$z_{LSRN6} \leq y_{SOLD}$
	$Z_{GSLN} \Rightarrow Y_{SOLD}$	$z_{GSLN} \leq y_{SOLD}$
	$Z_{S} \Rightarrow Y_{SOLD}$	$z_{S} \leq y_{SOLD}$
	$Z_{LPG5} \Rightarrow Y_{SOLD}$	$z_{LPG5} \leq y_{SOLD}$
	$Y_{SOLD} \Rightarrow Z_{SOLD}$	$y_{SOLD} \leq z_{SOLD}$
	$Z_{SOLD} \Rightarrow Y_{SOLD}$	$z_{SOLD} \leq y_{SOLD}$
BLND	$Y_{BLND} \Rightarrow Z_{ISO} \vee Z_{REF}$	$y_{BLND} \leq z_{ISO} + z_{REF}$
	$Z_{ISO} \Rightarrow Y_{BLND}$	$z_{ISO} \leq y_{BLND}$
	$Z_{REF} \Rightarrow Y_{BLND}$	$z_{REF} \leq y_{BLND}$
	$Y_{BLND} \Rightarrow Z_{GSLN}$	$y_{BLND} \leq z_{GSLN}$
	$Z_{GSLN} \Rightarrow Y_{BLND}$	$z_{GSLN} \leq y_{BLND}$
LPG	$Z_{LPG4} \Rightarrow Y_{LPG}$	$z_{LPG4} \leq y_{LPG}$
	$Y_{LPG} \Rightarrow Z_{LPG4}$	$y_{LPG} \leq z_{LPG4}$
	$Z_{LPG5} \Rightarrow Y_{LPG}$	$z_{LPG5} \leq y_{LPG}$
	$Y_{LPG} \Rightarrow Z_{LPG5}$	$y_{LPG} \leq z_{LPG5}$
FGH	$Y_{FGH} \Rightarrow Z_{FG1} \vee Z_{FG2} \vee Z_{FG3} \vee Z_{FG4}$	$y_{FGH} \leq z_{FG1} + z_{FG2} + z_{FG3} + z_{FG4}$
	$Z_{FG1} \Rightarrow Y_{FGH}$	$z_{FG1} \leq y_{FGH}$
	$Z_{FG2} \Rightarrow Y_{FGH}$	$z_{FG2} \leq y_{FGH}$
	$Z_{FG3} \Rightarrow Y_{FGH}$	$z_{FG3} \leq y_{FGH}$
	$Z_{FG4} \Rightarrow Y_{FGH}$	$z_{FG4} \leq y_{FGH}$
	$Y_{FGH} \Rightarrow Z_{FG5}$	$z_{FG5} \leq y_{FGH}$

(continued)

Table 5.4 (Continued)

Unit	Logic proposition	Algebraic constraint
SPLT1	$Z_{LSRN4} \Rightarrow Y_{SPLT1}$	$z_{LSRN4} \le y_{SPLT1}$
	$Y_{SPLT1} \Rightarrow Z_{LSRN4}$	$y_{SPLT1} \le z_{LSRN4}$
	$Y_{SPLT1} \Rightarrow Z_{LSRN5} \vee Z_{LSRN6}$	$y_{SPLT1} \le z_{LSRN5} + z_{LSRN6}$
	$Z_{LSRN5} \Rightarrow Y_{SPLT1}$	$z_{LSRN5} \le y_{SPLT1}$
	$Z_{LSRN6} \Rightarrow Y_{SPLT1}$	$z_{LSRN6} \le y_{SPLT1}$
SPLT2	$Z_{H2} \Rightarrow Y_{SPLT2}$	$z_{H2} \le y_{SPLT2}$
	$Y_{SPLT2} \Rightarrow Z_{H2}$	$y_{SPLT2} \le z_{H2}$
	$Y_{SPLT2} \Rightarrow Z_{H2\text{-}1} \vee Z_{H2\text{-}2}$	$y_{SPLT2} \le z_{H2\text{-}1} + z_{H2\text{-}1}$
	$Z_{H2\text{-}1} \Rightarrow Y_{SPLT2}$	$z_{H2\text{-}1} \le y_{SPLT2}$
	$Z_{H2\text{-}2} \Rightarrow Y_{SPLT2}$	$z_{H2\text{-}2} \le y_{SPLT2}$
MIX1	$Y_{MIX1} \Rightarrow Z_{HSRN1} \vee Z_{VIS1} \vee Z_{COK1} \vee Z_{FCC1} \vee Z_{HCR1} \vee Z_{PCHN1\text{-}1}$	$y_{MIX1} \le z_{HSRN1} + z_{VIS1} + z_{COK1} + z_{FCC1} + z_{HCR1} + z_{PCHN1\text{-}1}$
	$Z_{HSRN1} \Rightarrow Y_{MIX1}$	$z_{HSRN1} \le y_{MIX1}$
	$Z_{VIS1} \Rightarrow Y_{MIX1}$	$z_{VIS1} \le y_{MIX1}$
	$Z_{COK1} \Rightarrow Y_{MIX1}$	$z_{COK1} \le y_{MIX1}$
	$Z_{FCC1} \Rightarrow Y_{MIX1}$	$z_{FCC1} \le y_{MIX1}$
	$Z_{HCR1} \Rightarrow Y_{MIX1}$	$z_{HCR1} \le y_{MIX1}$
	$Z_{PCHN1\text{-}1} \Rightarrow Y_{MIX1}$	$z_{PCHN1\text{-}1} \le y_{MIX1}$
	$Y_{MIX1} \Rightarrow Z_{HSRN2}$	$y_{MIX1} \le z_{HSRN2}$
	$Z_{HSRN2} \Rightarrow Y_{MIX1}$	$z_{HSRN2} \le y_{MIX1}$
MIX2	$Y_{MIX2} \Rightarrow Z_{NAP1} \vee Z_{VIS2} \vee Z_{COK2} \vee Z_{FCC2} \vee Z_{HCR2} \vee Z_{PCHN1\text{-}2}$	$y_{MIX2} \le z_{NAP1} + z_{VIS2} + z_{COK2} + z_{FCC2} + z_{HCR2} + z_{PCHN1\text{-}2}$
	$Z_{NAP1} \Rightarrow Y_{MIX2}$	$z_{NAP1} \le y_{MIX2}$
	$Z_{VIS2} \Rightarrow Y_{MIX2}$	$z_{VIS2} \le y_{MIX2}$
	$Z_{COK2} \Rightarrow Y_{MIX2}$	$z_{COK2} \le y_{MIX2}$
	$Z_{FCC2} \Rightarrow Y_{MIX2}$	$z_{FCC2} \le y_{MIX2}$
	$Z_{HCR2} \Rightarrow Y_{MIX2}$	$z_{HCR2} \le y_{MIX2}$
	$Z_{PCHN1\text{-}2} \Rightarrow Y_{MIX2}$	$z_{PCHN1\text{-}2} \le y_{MIX2}$
	$Y_{MIX2} \Rightarrow Z_{NAP2}$	$y_{MIX2} \le z_{NAP2}$
	$Z_{NAP2} \Rightarrow Y_{MIX2}$	$z_{NAP2} \le y_{MIX2}$
MIX3	$Y_{MIX3} \Rightarrow Z_{LSRN1} \vee Z_{PCHN2} \vee Z_{LSRN2} \vee Z_{LSRN3}$	$y_{MIX3} \le z_{LSRN1} + z_{PCHN2} + z_{LSRN2} + z_{LSRN3}$
	$Z_{LSRN1} \Rightarrow Y_{MIX3}$	$z_{LSRN1} \le y_{MIX3}$
	$Z_{PCHN2} \Rightarrow Y_{MIX3}$	$z_{PCHN2} \le y_{MIX3}$
	$Z_{LSRN2} \Rightarrow Y_{MIX3}$	$z_{LSRN2} \le y_{MIX3}$
	$Z_{LSRN3} \Rightarrow Y_{MIX3}$	$z_{LSRN3} \le y_{MIX3}$
	$Y_{MIX3} \Rightarrow Z_{LSRN4}$	$y_{MIX3} \le z_{LSRN4}$
	$Z_{LSRN4} \Rightarrow Y_{MIX3}$	$z_{LSRN4} \le y_{MIX3}$

(continued)

Table 5.4 (Continued)

Unit	Logic proposition	Algebraic constraint
MIX4	$Y_{MIX4} \Rightarrow Z_{HSRN3} \vee Z_{HCR3} \vee Z_{PCHN3\text{-}1} \vee Z_{HSRN4}$	$y_{MIX4} \leq z_{HSRN3} + z_{HCR3} + z_{PCHN3\text{-}1} + z_{HSRN4}$
	$Z_{HSRN3} \Rightarrow Y_{MIX4}$	$z_{HSRN3} \leq y_{MIX4}$
	$Z_{HCR3} \Rightarrow Y_{MIX4}$	$z_{HCR} \leq y_{MIX4}$
	$Z_{PCHN3\text{-}1} \Rightarrow Y_{MIX4}$	$z_{PCHN3\text{-}1} \leq y_{MIX4}$
	$Z_{HSRN4} \Rightarrow Y_{MIX4}$	$z_{HSRN4} \leq y_{MIX4}$
	$Y_{MIX4} \Rightarrow Z_{HSRN5}$	$y_{MIX4} \leq z_{HSRN5}$
	$Z_{HSRN5} \Rightarrow Y_{MIX4}$	$z_{HSRN5} \leq y_{MIX4}$
MIX5	$Y_{MIX5} \Rightarrow Z_{NAP3} \vee Z_{NAP4} \vee Z_{HCR4} \vee Z_{PCHN3\text{-}2}$	$y_{MIX5} \leq z_{NAP3} + z_{NAP4} + z_{HCR4} + z_{PCHN3\text{-}2}$
	$Z_{NAP3} \Rightarrow Y_{MIX5}$	$z_{NAP3} \leq y_{MIX5}$
	$Z_{NAP4} \Rightarrow Y_{MIX5}$	$z_{NAP4} \leq y_{MIX5}$
	$Z_{HCR4} \Rightarrow Y_{MIX5}$	$z_{HCR4} \leq y_{MIX5}$
	$Z_{PCHN3\text{-}2} \Rightarrow Y_{MIX5}$	$z_{PCHN3\text{-}2} \leq y_{MIX5}$
	$Y_{MIX5} \Rightarrow Z_{NAP5}$	$y_{MIX5} \leq z_{NAP5}$
	$Z_{NAP5} \Rightarrow Y_{MIX5}$	$z_{NAP5} \leq y_{MIX5}$
MIX6	$Y_{MIX6} \Rightarrow Z_{LPG1} \vee Z_{LPG2} \vee Z_{LPG3}$	$y_{MIX6} \leq z_{LPG1} + z_{LPG2} + z_{LPG3}$
	$Z_{LPG1} \Rightarrow Y_{MIX6}$	$z_{LPG1} \leq y_{MIX6}$
	$Z_{LPG2} \Rightarrow Y_{MIX6}$	$z_{LPG2} \leq y_{MIX6}$
	$Z_{LPG3} \Rightarrow Y_{MIX6}$	$z_{LPG3} \leq y_{MIX6}$
	$Y_{MIX6} \Rightarrow Z_{LPG4}$	$y_{MIX6} \leq z_{LPG4}$
	$Z_{LPG4} \Rightarrow Y_{MIX6}$	$z_{LPG4} \leq y_{MIX6}$
VIS	$Y_{VIS} \Rightarrow Z_{VIS1} \vee Z_{VIS2}$	$y_{VIS} \leq z_{VIS1} + z_{VIS2}$
	$Z_{VIS1} \Rightarrow Y_{VIS}$	$z_{VIS1} \leq y_{VIS}$
	$Z_{VIS2} \Rightarrow Y_{VIS}$	$z_{VIS2} \leq y_{VIS}$
COK	$Y_{COK} \Rightarrow Z_{COK1} \vee Z_{COK2}$	$y_{COK} \leq z_{COK1} + z_{COK2}$
	$Z_{COK1} \Rightarrow Y_{COK}$	$z_{COK1} \leq y_{COK}$
	$Z_{COK2} \Rightarrow Y_{COK}$	$z_{COK2} \leq y_{COK}$
FCC	$Y_{FCC} \Rightarrow Z_{FCC1} \vee Z_{FCC2}$	$y_{FCC} \leq z_{FCC1} + z_{FCC2}$
	$Z_{FCC1} \Rightarrow Y_{FCC}$	$z_{FCC1} \leq y_{FCC}$
	$Z_{FCC2} \Rightarrow Y_{FCC}$	$z_{FCC2} \leq y_{FCC}$
HCR	$Y_{HCR} \Rightarrow Z_{HCR1} \vee Z_{HCR2} \vee Z_{HCR3} \vee Z_{HCR4}$	$y_{HCR} \leq z_{HCR1} + z_{HCR2} + z_{HCR3} + z_{HCR4}$
	$Z_{HCR1} \Rightarrow Y_{HCR}$	$z_{HCR1} \leq y_{HCR}$
	$Z_{HCR2} \Rightarrow Y_{HCR}$	$z_{HCR2} \leq y_{HCR}$
	$Z_{HCR3} \Rightarrow Y_{HCR}$	$z_{HCR3} \leq y_{HCR}$
	$Z_{HCR4} \Rightarrow Y_{HCR}$	$z_{HCR4} \leq y_{HCR}$

On the other hand, Approach 2 involves formulation by employing logical constraints on design specifications (as shown in the following for ADU):

$$Y_{\mathrm{ADU}} \Leftrightarrow \left(Y_{\mathrm{HDT1}} \vee Y_{\mathrm{HDT2}}\right)$$

$$\neg Y_{\mathrm{ADU}} \vee \left(Y_{\mathrm{HDT1}} \vee Y_{\mathrm{HDT2}}\right)$$

$$\begin{bmatrix} \neg Y_{\mathrm{ADU}} \\ f_{\mathrm{NAP1}} = 0 \\ f_{\mathrm{LSRN1}} = 0 \\ f_{\mathrm{HSRN1}} = 0 \\ c_{\mathrm{ADU}} = 0 \end{bmatrix} \vee \begin{bmatrix} Y_{\mathrm{HDT1}} \\ 0.0109 \rightleftharpoons \left(f_{\mathrm{H2}} + f_{\mathrm{HSRN2}}\right) = f_{\mathrm{FG1}} \\ 0.0012 \rightleftharpoons \left(f_{\mathrm{H2}} + f_{\mathrm{HSRN2}}\right) = f_{\mathrm{H2S1}} \\ 0.0058 \rightleftharpoons \left(f_{\mathrm{H2}} + f_{\mathrm{HSRN2}}\right) = f_{\mathrm{LPG}\rightleftharpoons 1} \\ 0.2610 \rightleftharpoons \left(f_{\mathrm{H2}} + f_{\mathrm{HSRN}\rightleftharpoons 2}\right) = f_{\mathrm{LSRN}\rightleftharpoons 2} \\ 0.7211 \rightleftharpoons \left(f_{\mathrm{H2}} + f_{\mathrm{HSRN}\rightleftharpoons 2}\right) = f_{\mathrm{HSRN3}} \\ 0.9821 \rightleftharpoons \left(f_{\mathrm{H2}} + f_{\mathrm{HSRN}\rightleftharpoons 2}\right) = f_{\mathrm{NAP4}} \\ c_{\mathrm{HDT1}} = 96 \end{bmatrix} \vee$$

$$\begin{bmatrix} Y_{\mathrm{HDT2}} \\ 0.0109 \rightleftharpoons \left(f_{\mathrm{H2}} + f_{\mathrm{NAP2}}\right) = f_{\mathrm{FG2}} \\ 0.0012 \rightleftharpoons \left(f_{\mathrm{H2}} + f_{\mathrm{NAP2}}\right) = f_{\mathrm{H2S2}} \\ 0.0058 \rightleftharpoons \left(f_{\mathrm{H2}} + f_{\mathrm{NAP2}}\right) = f_{\mathrm{LPG3}} \\ 0.2610 \rightleftharpoons \left(f_{\mathrm{H2}} + f_{\mathrm{NAP2}}\right) = f_{\mathrm{LSRN3}} \\ 0.7211 \rightleftharpoons \left(f_{\mathrm{H2}} + f_{\mathrm{NAP2}}\right) = f_{\mathrm{HSRN4}} \\ 0.9821 \rightleftharpoons \left(f_{\mathrm{H2}} + f_{\mathrm{NAP2}}\right) = f_{\mathrm{NAP3}} \\ c_{\mathrm{HDT2}} = 96 \end{bmatrix} \tag{5.33}$$

In this work, Approach 1 is applied to the entire superstructure with the complete formulation as listed in Table 5.5.

5.4 Numerical Implementation for Computational Experiments

Based on the established ease and advantages of using general algebraic modeling system (GAMS) discussed in Chapter 4, we can perform numerical studies for both deterministic and stochastic formulations of models by coding and implementing them using GAMS integrated development environment (IDE). The models are solved using an appropriate solver such as CPLEX (IBM 2020) or Gurobi (Gurobi Optimization 2020) for the linear cases or instances and CONOPT (Drud 2021) for those that are nonlinear. CPLEX is the default solver in GAMS for handling linear optimization models, employing an algorithm based on interior point methods (Karmarkar 1984). For nonlinear optimization models, the default GAMS solver is CONOPT, which is based on a feasible path generalized reduced gradient method with restoration (Drud 2021).

The solution generated for the deterministic equivalent formulation of a stochastic optimization model consists of (i) first-stage decision variables of production flow rates for all process streams; and (ii) second-stage recourse variables of production deviations due to randomness (uncertainty) in parameters such as market demands for products (which are right-hand-side constraint coefficients) and process unit yields (which are left-hand-side constraint coefficients). If a stochastic model is

Table 5.5 GDP direct formulation on process unit existence based on component material balances.

Unit	Disjunction	
ADU	$\begin{bmatrix} Y_{ADU} \\ (0.2088) \rightleftharpoons f_{CR} = f_{NAP1} \\ (0.0555) \rightleftharpoons f_{CR} = f_{LSRN1} \\ (0.1533) \rightleftharpoons f_{CR} = f_{HSRN1} \\ f_{NAP1} = f_{LSRN1} + f_{HSRN1} \\ c_{ADU} = 228 \end{bmatrix} \vee \begin{bmatrix} \neg Y_{ADU} \\ f_{CR} = 0 \\ f_{NAP1} = 0 \\ f_{LSRN1} = 0 \\ f_{HSRN1} = 0 \\ c_{ADU} = 0 \end{bmatrix}$	(5.36)
HDT1	$\begin{bmatrix} Y_{HDT1} \\ 0.0109 \rightleftharpoons (f_{H2} + f_{HSRN2}) = f_{FG1} \\ 0.0012 \rightleftharpoons (f_{H2} + f_{HSRN2}) = f_{H2S1} \\ 0.0058 \rightleftharpoons (f_{H2} + f_{HSRN2}) = f_{LPG1} \\ 0.2610 \rightleftharpoons (f_{H2} + f_{HSRN2}) = f_{LSRN2} \\ 0.7211 \rightleftharpoons (f_{H2} + f_{HSRN2}) = f_{HSRN3} \\ 0.9821 \rightleftharpoons (f_{H2} + f_{HSRN2}) = f_{NAP4} \\ c_{HDT1} = 96 \end{bmatrix} \vee \begin{bmatrix} \neg Y_{HDT1} \\ f_{H2} = f_{HSRN2} = 0 \\ f_{FG1} = 0 \\ f_{H2S1} = 0 \\ f_{LPG1} = 0 \\ f_{LSRN2} = 0 \\ f_{HSRN3} = 0 \\ f_{NAP4} = 0 \\ c_{HDT\,1} = 0 \end{bmatrix}$	(5.37)
HDT2	$\begin{bmatrix} Y_{HDT2} \\ 0.0109\,(f_{H2} + f_{NAP2}) = f_{FG2} \\ 0.0012\,(f_{H2} + f_{NAP2}) = f_{H2S2} \\ 0.0058\,(f_{H2} + f_{NAP2}) = f_{LPG3} \\ 0.2610\,(f_{H2} + f_{NAP2}) = f_{LSRN3} \\ 0.7211\,(f_{H2} + f_{NAP2}) = f_{HSRN4} \\ 0.9821\,(f_{H2} + f_{NAP2}) = f_{NAP3} \\ c_{HDT2} = 96 \end{bmatrix} \vee \begin{bmatrix} \neg Y_{HDT2} \\ f_{H2} = f_{NAP2} = 0 \\ f_{FG2} = 0 \\ f_{H2S2} = 0 \\ f_{LPG3} = 0 \\ f_{LSRN3} = 0 \\ f_{HSRN4} = 0 \\ f_{NAP3} = 0 \\ c_{HDT2} = 0 \end{bmatrix}$	(5.38)
ISO	$\begin{bmatrix} Y_{ISO} \\ (0.99) f_{LSRN5} \rightleftharpoons\rightleftharpoons= f_{ISO} \\ (0.01) f_{LSRN5} = f_{FG4} \\ c_{ISO} = 42 \end{bmatrix} \vee \begin{bmatrix} \neg Y_{ISO} \\ f_{LSRN\rightleftharpoons 5} = 0 \\ f_{ISO} = 0 \\ f_{FG4} = 0 \\ c_{ISO} = 0 \end{bmatrix}$	(5.39)
SRU	$\begin{bmatrix} Y_{SRU} \\ 0.8478\,(f_{H2S1} + \rightleftharpoons f_{H2S2}) = f_{S\rightleftharpoons} \\ 0.1522\,(f_{H2S1} + \rightleftharpoons f_{H2S2}) = f_{TG\rightleftharpoons} \\ c_{SRU} = 30 \end{bmatrix} \vee \begin{bmatrix} \neg Y_{SRU} \\ f_{H2S1} = f_{H2S2} = 0 \\ f_S = 0 \\ f_{TG} = 0 \\ c_{SRU} = 0 \end{bmatrix}$	(5.40)

(continued)

Table 5.5 (Continued)

Unit	Disjunction	
REF	$\begin{bmatrix} Y_{REF} \\ 0.0320\left(f_{HSRN5}+f_{NAP5}\right)=f_{H2} \\ 0.0370\left(f_{HSRN5}+f_{NAP5}\right)=f_{FG3} \\ 0.0780\left(f_{HSRN5}+f_{NAP5}\right)=f_{LPG2} \\ 0.8530\left(f_{HSRN5}+f_{NAP5}\right)=f_{REF\rightleftharpoons} \\ c_{REF}=270 \end{bmatrix} \vee \begin{bmatrix} \neg Y_{REF} \\ f_{HSRN5}=f_{NAP5}=0 \\ f_{H2}=0 \\ f_{FG3}=0 \\ f_{LPG2}=0 \\ f_{REF}=0 \\ c_{REF}=0 \end{bmatrix}$	(5.41)
SOLD	$\begin{bmatrix} Y_{SOLD} \\ f_{SOLD}=f_{LSRN6}+f_{S}+f_{GSLN}+f_{LPG5} \\ c_{SOLD}=10 \end{bmatrix} \vee \begin{bmatrix} \neg Y_{SOLD} \\ f_{SOLD}=f_{LSRN6}=f_{S}=f_{GSLN}=f_{LPG5}=0 \\ c_{SOLD}=0 \end{bmatrix}$	(5.42)
BLND	$\begin{bmatrix} Y_{GSLN} \\ f_{GSLN}=f_{ISO}+f_{REF} \\ c_{GSLN}=10 \end{bmatrix} \vee \begin{bmatrix} \neg Y_{GSLN} \\ f_{GSLN}=f_{ISO}=f_{REF}=0 \\ c_{GSLN}=0 \end{bmatrix}$	(5.43)
LPG	$\begin{bmatrix} Y_{LPG} \\ f_{LPG5}=f_{LPG4} \\ c_{LPG}=10 \end{bmatrix} \vee \begin{bmatrix} \neg Y_{LPG} \\ f_{LPG5}=f_{LPG4}=0 \\ c_{LPG}=0 \end{bmatrix}$	(5.44)
FGH	$\begin{bmatrix} Y_{FG} \\ f_{FG5}=f_{FG1}+f_{FG2}+f_{FG3}+f_{FG4} \\ c_{FG}=10 \end{bmatrix} \vee \begin{bmatrix} \neg Y_{FG} \\ f_{FG5}=f_{FG1}=f_{FG2}=f_{FG3}=f_{FG4}=0 \\ c_{FG}=0 \end{bmatrix}$	(5.45)
10SPLT1	$\begin{bmatrix} Y_{SPLT1} \\ (0.9)f_{LSRN4}\rightleftharpoons=f_{LSRN5} \\ (0.1)f_{LSRN4}\rightleftharpoons=f_{LSRN6} \\ c_{SPLT1}=\alpha \end{bmatrix} \vee \begin{bmatrix} \neg Y_{SPLT1} \\ f_{LSRN4}=f_{LSRN5}=f_{LSRN6}=0 \\ c_{SPLT1}=0 \end{bmatrix}$	(5.46)

(continued)

Table 5.5 (Continued)

Unit	Disjunction
SPLT2	$$\begin{bmatrix} Y_{\text{SPLT2}} \\ f_{\text{H2}} \rightleftharpoons = f_{\text{H2\_1}} + f_{\text{H2\_2}} \\ c_{\text{SPLT2}} = 10 \end{bmatrix} \vee \begin{bmatrix} \neg Y_{\text{SPLT2}} \\ f_{\text{H2}} \rightleftharpoons = f_{\text{H2\_1}} = f_{\text{H2\_2}} = 0 \\ c_{\text{SPLT2}} = 0 \end{bmatrix} \quad (5.47)$$
MIX1	$$\begin{bmatrix} Y_{\text{MIX1}} \\ f_{\text{NAP1}} + f_{\text{VIS\_2}} + f_{\text{COK\_2}} + f_{\text{FCC\_2}} + f_{\text{HCR\_2}} + f_{\text{PCHN1\_2}} \rightleftharpoons = f_{\text{NAP2}} \\ c_{\text{MIX1}} = 10 \end{bmatrix} \vee \begin{bmatrix} \neg Y_{\text{MIX1}} \\ f_{\text{NAP1}} = f_{\text{VIS\_2}} = f_{\text{COK\_2}} = f_{\text{FCC\_2}} = 0 \\ f_{\text{HCR\_2}} = f_{\text{PCHN1\_2}} \rightleftharpoons = f_{\text{NAP2}} = 0 \\ c_{\text{MIX1}} = 0 \end{bmatrix} \quad (5.48)$$
MIX2	$$\begin{bmatrix} Y_{\text{MIX2}} \\ f_{\text{HSRN1}} + f_{\text{VIS\_1}} + f_{\text{COK\_1}} + f_{\text{FCC\_1}} + f_{\text{HCR\_1}} \\ + f_{\text{PCHN1\_1}} \rightleftharpoons = f_{\text{HSRN2}} \\ c_{\text{MIX2}} = 10 \end{bmatrix} \vee \begin{bmatrix} \neg Y_{\text{MIX2}} \\ f_{\text{HSRN1}} = f_{\text{VIS\_1}} = f_{\text{COK\_1}} = f_{\text{FCC\_1}} = 0 \\ f_{\text{HCR\_1}} = f_{\text{PCHN1\_1}} \rightleftharpoons = f_{\text{HSRN2}} = 0 \\ c_{\text{MIX2}} = 0 \end{bmatrix} \quad (5.49)$$
MIX3	$$\begin{bmatrix} Y_{\text{MIX3}} \\ f_{\text{LSRN1}} + f_{\text{LSRN2}} + f_{\text{LSRN3}} + f_{\text{PCHN2}} = f_{\text{LSRN4}} \\ c_{\text{MIX3}} = 10 \end{bmatrix} \vee \begin{bmatrix} \neg Y_{\text{MIX3}} \\ f_{\text{LSRN1}} = f_{\text{LSRN2}} = f_{\text{LSRN3}} = f_{\text{PCHN2}} = f_{\text{LSRN4}} = 0 \\ c_{\text{MIX3}} = 0 \end{bmatrix} \quad (5.50)$$
MIX4	$$\begin{bmatrix} Y_{\text{MIX4}} \\ f_{\text{HSRN3}} + f_{\text{HSRN4}} + f_{\text{PCHN3\_1}} + f_{\text{HCR\_3}} = f_{\text{HSRN5}} \\ c_{\text{MIX4}} = 10 \end{bmatrix} \vee \begin{bmatrix} \neg Y_{\text{MIX4}} \\ f_{\text{HSRN3}} = f_{\text{HSRN4}} = f_{\text{PCHN3\_1}} = f_{\text{HCR\_3}} = f_{\text{HSRN5}} = 0 \\ c_{\text{MIX4}} = 0 \end{bmatrix} \quad (5.51)$$

(continued)

Table 5.5 (Continued)

Unit	Disjunction
MIX5	$\begin{bmatrix} Y_{\text{MIX5}} \\ f_{\text{NAP3}} + f_{\text{NAP4}} + f_{\text{PCHN3\_2}} + f_{\text{HCR\_4}} = f_{\text{NAP5}} \\ c_{\text{MIX5}} = 10 \end{bmatrix} \vee \begin{bmatrix} \neg Y_{\text{MIX5}} \\ f_{\text{NAP3}} = f_{\text{NAP4}} = f_{\text{PCHN3\_2}} = f_{\text{HCR\_4}} = f_{\text{NAP5}} = 0 \\ c_{\text{MIX5}} = 0 \end{bmatrix}$ (5.52)
MIX6	$\begin{bmatrix} Y_{\text{MIX6}} \\ f_{\text{LPG1}} + f_{\text{LPG2}} + f_{\text{LPG3}} = f_{\text{LPG4}} \\ c_{\text{MIX6}} = 10 \end{bmatrix} \vee \begin{bmatrix} \neg Y_{\text{MIX6}} \\ f_{\text{LPG1}} = f_{\text{LPG2}} = f_{\text{LPG3}} = f_{\text{LPG4}} = 0 \\ c_{\text{MIX6}} = 0 \end{bmatrix}$ (5.53)

nonconvex and nonlinear, starting values for the first-stage decision variables can be initialized to an optimal solution obtained from the deterministic formulation toward attaining a globally optimal solution. Although a global optimum might not be guaranteed, not detecting multiple local solutions under such varied initial conditions can be considered to indicate that we have computed a global optimum.

In addition, it may be useful to note that nonlinear optimization algorithms often search in the space defined by superbasic variables (i.e. variables that are not in the basis but whose values are between the upper and lower bounds). If an infeasible solution is found by the solver used in GAMS, it is most likely due to the fact that a superbasic variable is not present for which the facility within GAMS readily reports.

5.5 Computational Experiment Examples

Numerical studies in this work are coded and implemented using GAMS 24.2.3. Numerical examples are solved using GAMS/CPLEX 10 (IBM 2019) for MILP problems and GAMS/CONOPT for NLP problems on a notebook computer running on an Intel processor with 2.0 GHz processing speed and 4.0 GB of memory (RAM).

The computational results discussion is divided into two parts, i.e. that for MILP and GDP. The associated model sizes and computational statistics are reported in Table 5.6. GAMS modeling code and computational outputs are given in the

Table 5.6 Model sizes and computational statistics.

Model type	**MILP**	**GDP**
Solver	**GAMS/CPLEX**	**GAMS/LogMIP**
Number of constraints	336	195
Number of continuous variables	77	95
Number of binary variables	80	22
Number of iterations	26	23
CPU time (s)	0.053	0.037

appendices. The MILP model is solved by using GAMS/CPLEX, which requires slightly more resource use of 0.053 s of CPU time as compared to that of 0.037 s for the GDP model solved by using GAMS/LogMIP, but the solution time for both models is considerably trivial and the difference insignificant.

In comparing model statistics, the MILP formulation has more equations and binary variables; hence, it requires more iterations for solution convergence as compared to the GDP formulation. This shows an advantage of GDP of implicitly modeling the binary decisions, which reduces the model size by only considering disjunctions for which the Boolean variables are true, thus obviating demanding combinatorial search effort by the binary variables (Yeomans and Grossmann 1999).

For each model type, two design scenarios as distinguished by API gravity of the crude oil charge to ADU are considered: light crude oil (API greater than 33) and heavy crude oil (API less than or equal to 33) processing. Furthermore, two variations are considered under each scenario, i.e. in terms of feed requirement and operating cost. The corresponding model to be solved is stipulated through the use of if–else selection statement in the GAMS implementation. Tables 5.7 and 5.8 list the base economic and operating parameter values, respectively. Table 5.9 gives the base operating cost of utilities, and Table 5.10 gives the base capital cost and utility requirements of major unit operations.

Table 5.7 Base economic parameters.

Parameter	**Value**
Annual operating time	330 day/year
Product demand (gasoline)	100,000 kg/day (lower bound)
Crude oil cost	120.0 RM/bbl
Naphtha cost	0.524 RM/kg
Nelson–Farrar Refinery Construction Index (NFRCI)	Jan 1991 = 1241.7, Dec 2008 = 2067.2

Source: Adapted from Maples (2000).

Table 5.8 Base operating parameters.

Parameter	Unit	Lower bound	Upper bound
Crude oil feed charge	barrel/day	100,000	150,000
Process unit capacity	kg/day	(vary according to specific unit)	1×10^8 (general value)

Source: Adapted from Maples (2000).

Table 5.9 Base operating cost on utilities (Malaysian Industrial Development Authority (MIDA) 2008).

Type	Unit of measure	Value
Electricity	RM per kWh	0.1980
Fuel	RM per MJ	0.1018
HP Steam	RM per kg	0.0050
Cooling Water	RM per m^3	0.8400

Source: Adapted from MIDA (2000).

Table 5.10 Base capital cost and utility requirements of process units for light crude oil processing (Maples 2000).

			Utility requirement			
Process	Jan 91 ($\times 10^6$ RM)	Dec 08 ($\times 10^6$ RM)	Electricity (MWh/kg)	Fuel (kJ/kg)	Steam (kg/kg)	Cooling water (m^3/kg)
ADU	137	228	0.0039	0.0826	0.0888	0.0000
VIS	86	144	0.0039	0.0660	0.1776	0.0000
COK	166	276	0.0282	0.0991	0.1421	0.0000
FCC	310	515	0.0078	0.0660	0.0710	0.0119
HCR	342	569	0.1402	0.2766	0.0000	0.0000
HDT	58	96	0.0157	0.0248	0.0533	0.0000
REF	162	270	0.0078	0.2477	0.1421	0.0030
ISO	25	42	0.0078	0.0083	0.1279	0.0000
SRU (per ton)	18	30	0.3132	0.0000	2.6636	0.1482

Source: Adapted from Maples (2000).

The following relations hold for the refinery economics calculation:

$$\text{Total capital investment} = \text{Fixed capital investment} + \text{Working capital,} \tag{5.34}$$

$$\text{Fixed capital investment} = \text{Total equipment base cost,} \tag{5.35}$$

$$\text{Total operating cost} = \text{Fixed and variable operating costs} + \text{General expenses,} \tag{5.36}$$

An objective function as adopted in this work is expressed by:

$$\text{Total cost} = \text{Total capital investment} + \text{Total operating cost.} \tag{5.37}$$

5.5.1 MILP Model Results

The model assumptions for considering feed requirement variation case are as follows:

- total unit input flow rates depend on light or heavy crude oil processing with that from VIS, COK, FCC, and HCR set at 2×10^6 kg/day,
- constant production requirements;
- 330 working days annually.

The model assumptions for considering the operating cost variation case are similar except that input flow rates from VIS, COK, and HCR are set at 2×10^6 kg/day (as similar to the previous case) while that from FCC is 3×10^6 kg/day.

Results from the MILP model reveal that the operating cost for heavy crude oil charge processing is more than that of light crude under varying feed and operating costs. The result is validated for heavy crude oil charge in which COK, FCC, and HCR are selected to perform more severe processing whereas VIS and FCC are selected for light crude oil charge processing. The operating costs for COK and HCR are higher than that of FCC and VIS, thus heavy crude oil charge processing cost is higher as tabulated in Tables 5.11 and 5.12. The objective function trend can be shown to agree with that for a typical refinery (Kovac et al. 2006).

Table 5.11 MILP main results summary for feed requirement variation case.

	Light crude oil	Heavy crude oil
CAPEX + OPEX + Raw material (mil RM)	2744	2743
CAPEX (mil RM)	791	791
OPEX + Raw material (mil RM)	1953	1951
Crude oil feed requirement (kg/day)	4.5E+7	3.2E+7
Raw material (mil RM)	42	30
OPEX (mil RM)	1910	1921
OPEX (mil RM/day)	5.7	5.8

Table 5.12 MILP main results summary for operating cost variation case.

	Light crude oil	Heavy crude oil
CAPEX + OPEX +Raw material (mil RM)	2688	2743
CAPEX (mil RM)	791	791
OPEX + Raw material (mil RM)	1897	1951
Crude feed requirement (kg/day)	3.28E+7	3.28E+7
Raw material (mil RM)	30	30
OPEX (mil RM)	1866	1921
OPEX (mil RM/day)	5.6	5.8

Figures 5.2 and 5.3 show the superstructures for light and heavy crude oil charges, respectively, with a shaded square indicating that the unit is selected and a shaded circle that the state is selected with an associated nonzero flow rate. The corresponding optimal configuration flowsheets are presented in Figures 5.4 and 5.5 for the respective crude oil charge scenarios. The optimal configurations obtained are generally in good agreement with a typical refinery topology.

5.5.2 GDP Model Results

The model assumptions are as follows: constant total unit input flow rates, which are independent of light or heavy crude oil with those from VIS, COK, and HCR set at 2×10^6 kg/day (as is similar to the previous case) while that from FCC is 3×10^6 kg/day, constant production requirements, 330 working days annually.

The main results summary in Table 5.13 indicates that the GDP model is in agreement with the MILP model in that OPEX for heavy crude oil charge processing is more than that of light crude oil. The total cost objective function value is also within the range of that reported for a typical refinery (Hydrocarbon Processing magazine 2006).

Superstructures indicating the feasible alternatives for naphtha produced from the atmospheric distillation unit as constituted by the selected units (in shaded squares) and streams (in shaded circles) for light and heavy crude oil charge scenarios are shown by the superstructures in Figures 5.6 and 5.7, respectively. Their corresponding optimal configurations or topologies are shown in Figures 5.8 and 5.9.

The optimal refinery topologies generated for the MILP and GDP models agree with typical existing refinery topology (Hydrocarbon Processing magazine 2006) as based on validating the objective function value of minimum cost against a set of industrial data (Gary et al. 2007). An observed difference between the two models is in terms of the less CPU time or resource usage requirement of GDP as compared to that of MILP, which can be attributed to the fact that GDP avoids the use of big-M constraints that have been reported to yield poor relaxations (Turkay and Grossmann 1996).

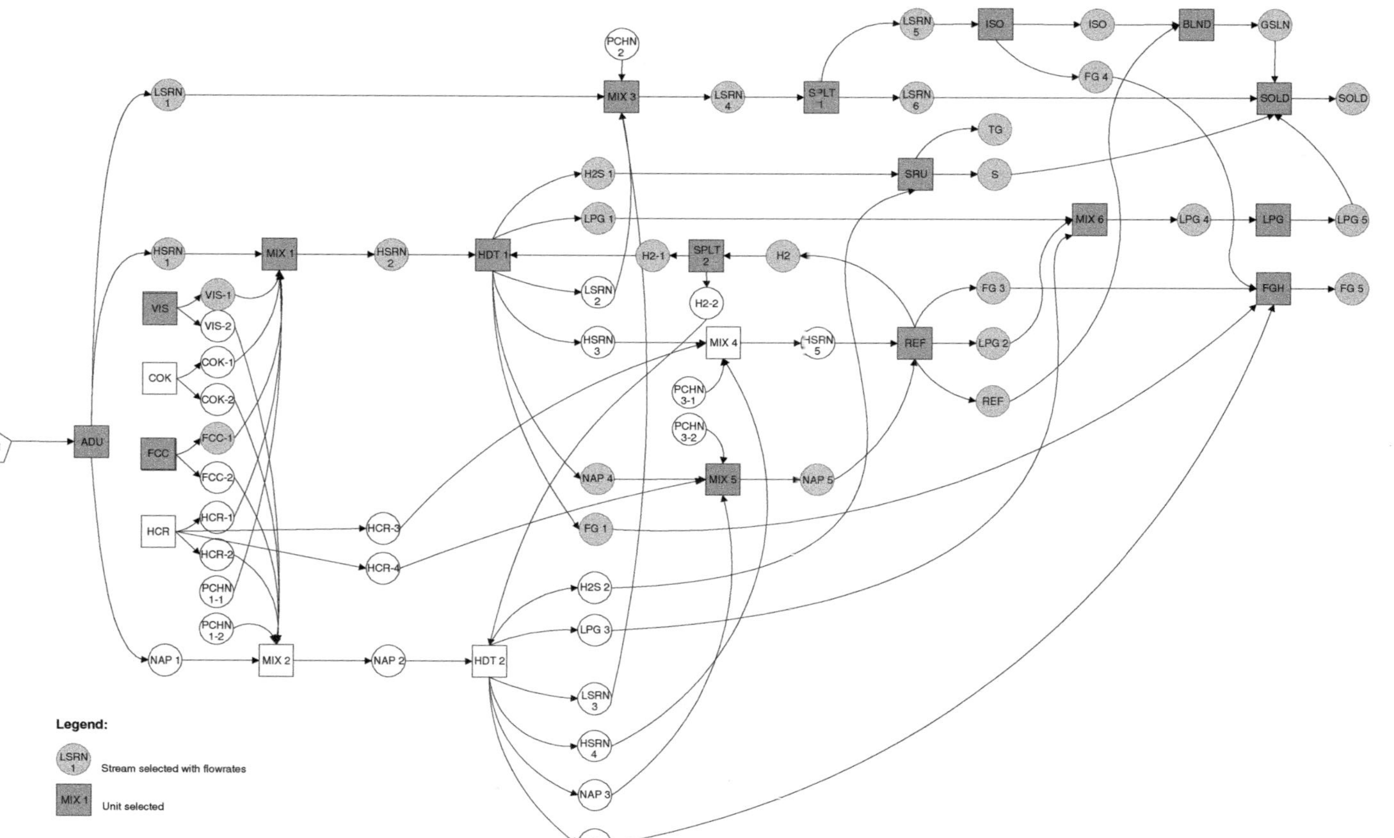

Figure 5.2 MILP result on superstructure with optimally selected units indicated for naphtha production with light crude oil charge (similar for both feed requirement and operating cost variation cases).

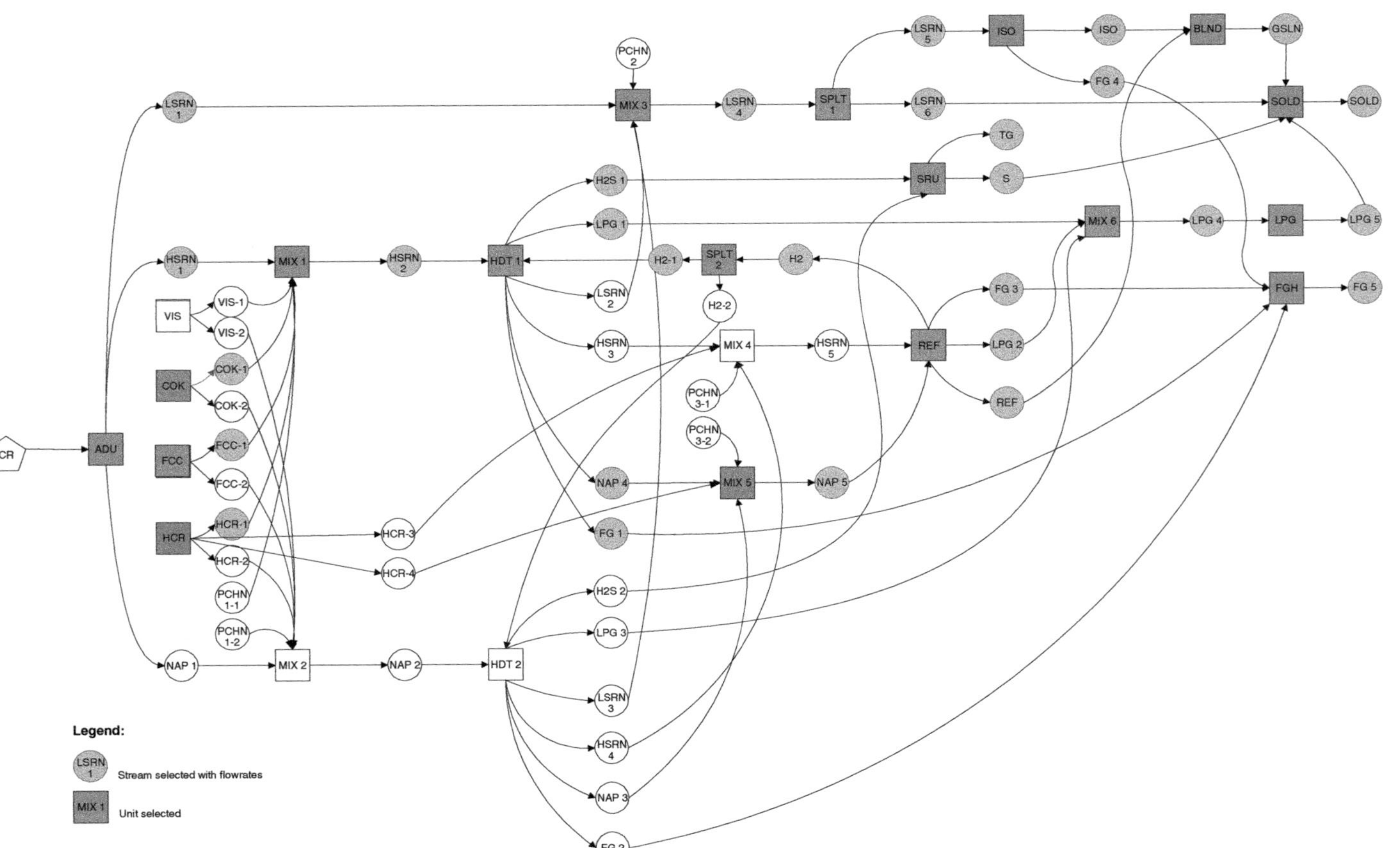

Figure 5.3 MILP result on superstructure with optimally selected units indicated for naphtha production with heavy crude oil charge.

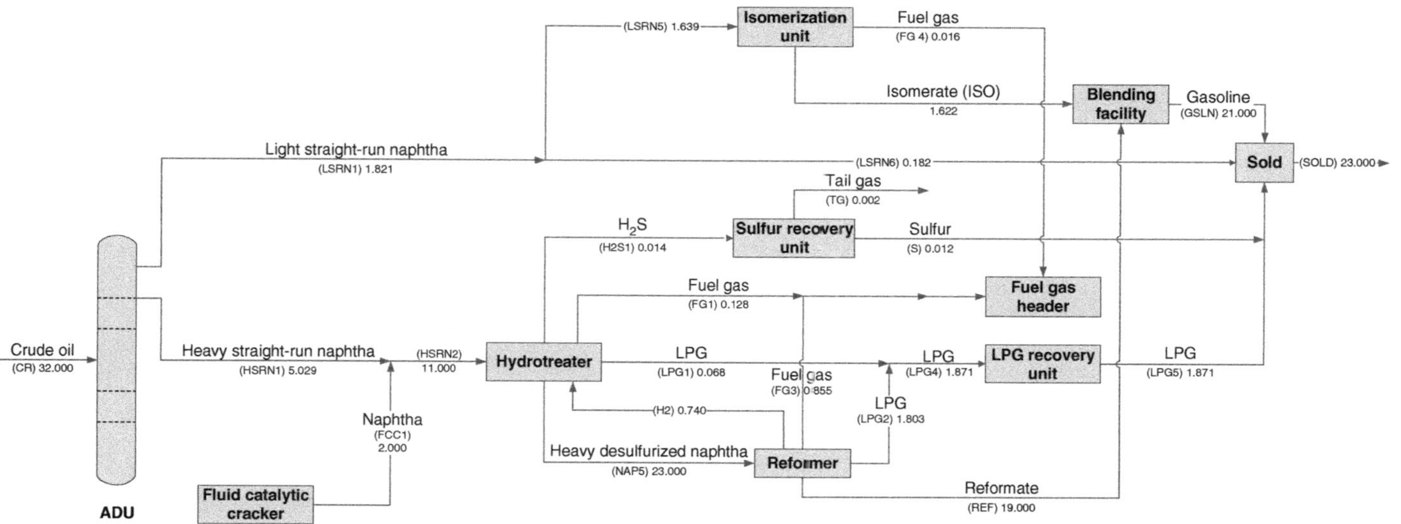

Figure 5.4 MILP result on optimal naphtha production configuration for light crude oil charge.

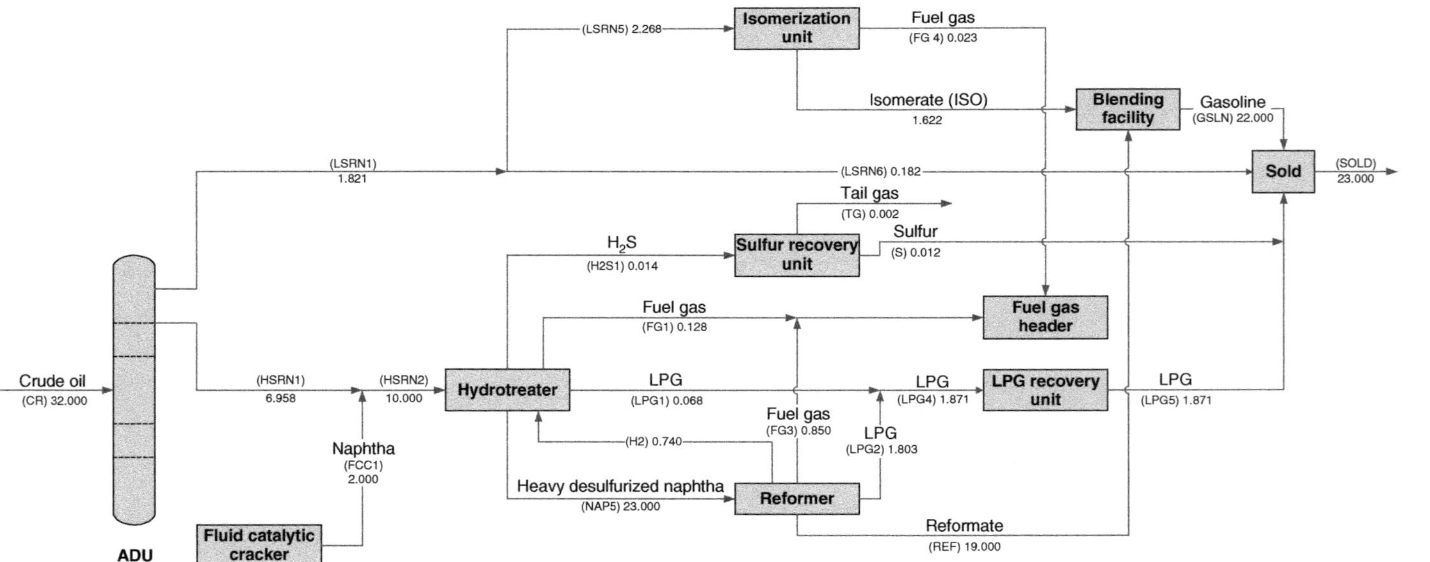

Figure 5.5 MILP result on optimal naphtha production configuration for heavy crude oil charge.

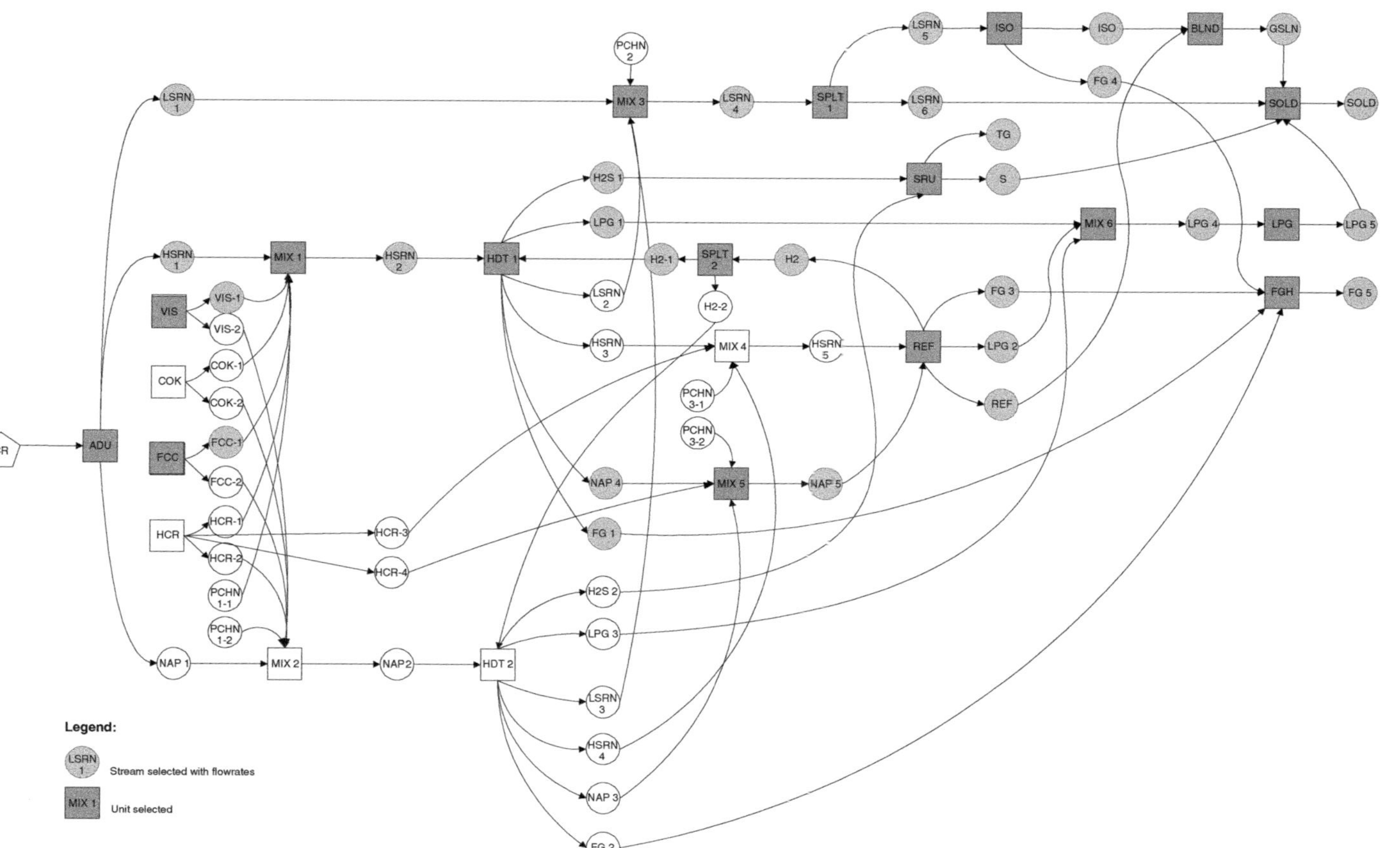

Figure 5.6 MILP result on superstructure with optimally selected units indicated for napntha production with light crude oil charge.

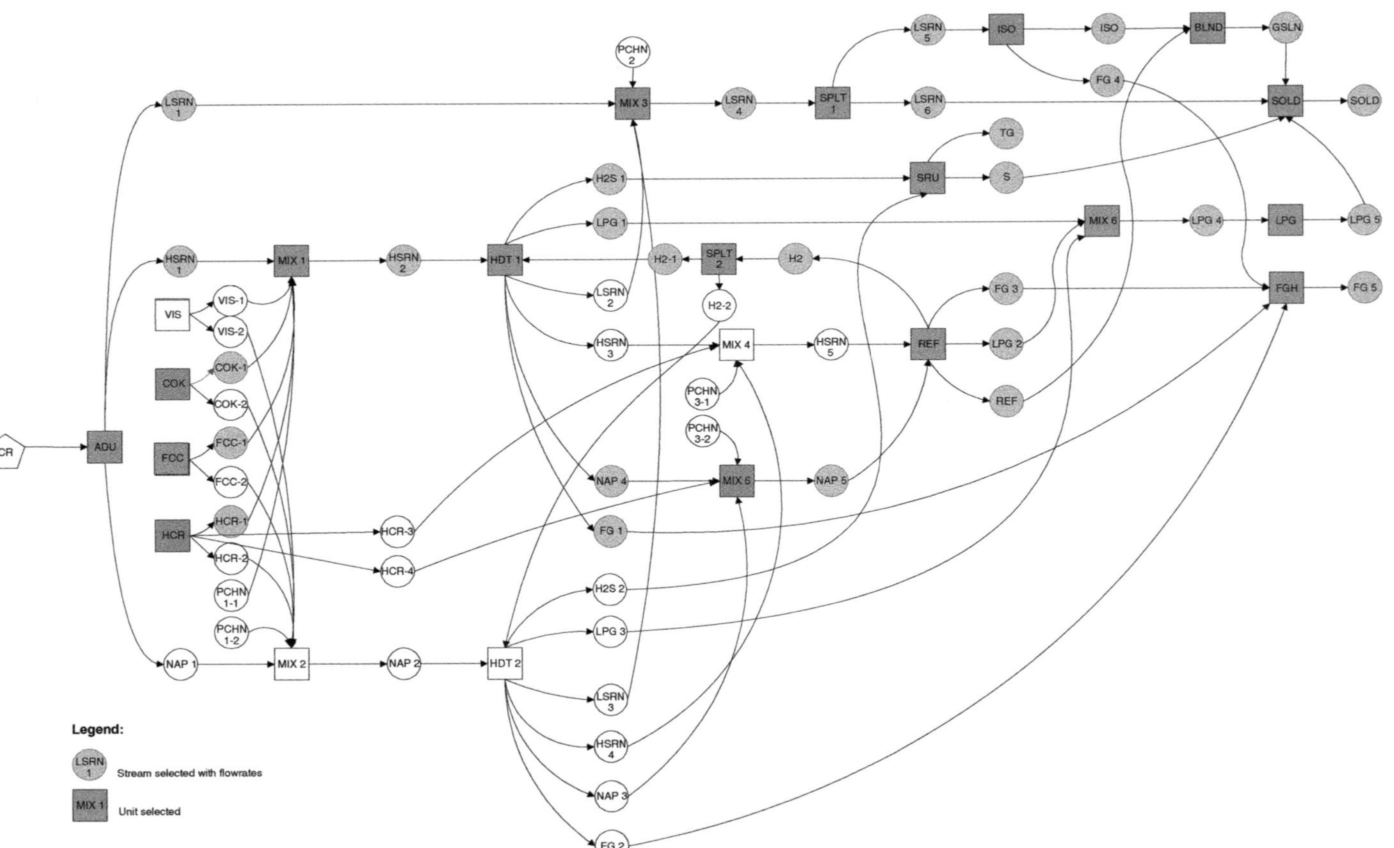

Figure 5.7 MILP result on superstructure with optimally selected units indicated for naphtha production with heavy crude oil charge.

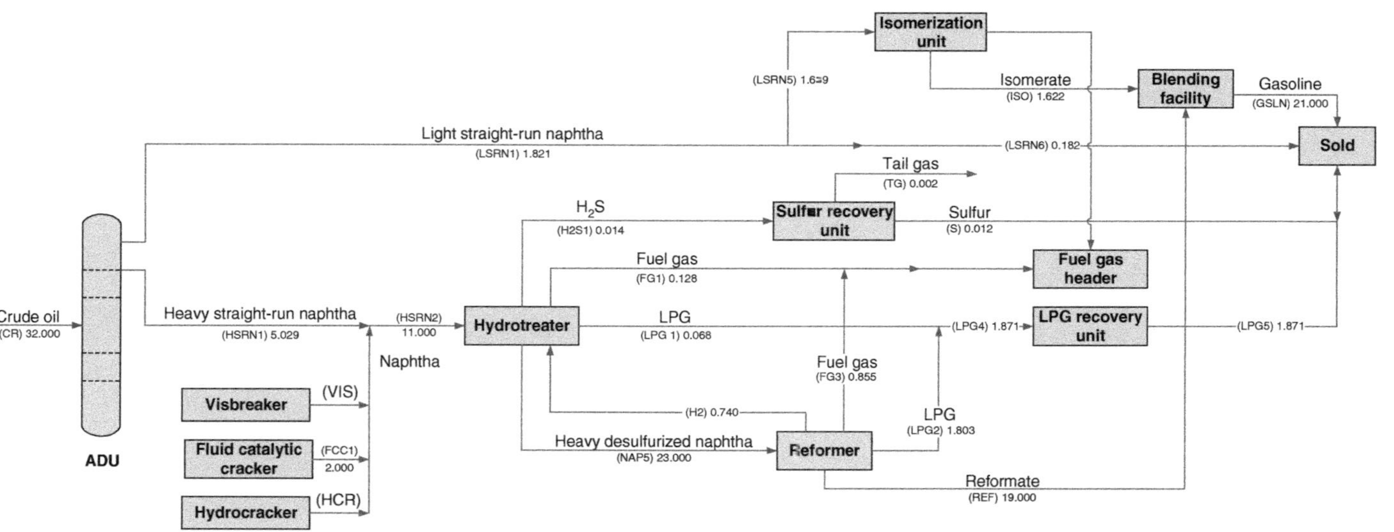

Figure 5.8 GDP result on optimal naphtha production configuration for light crude oil charge.

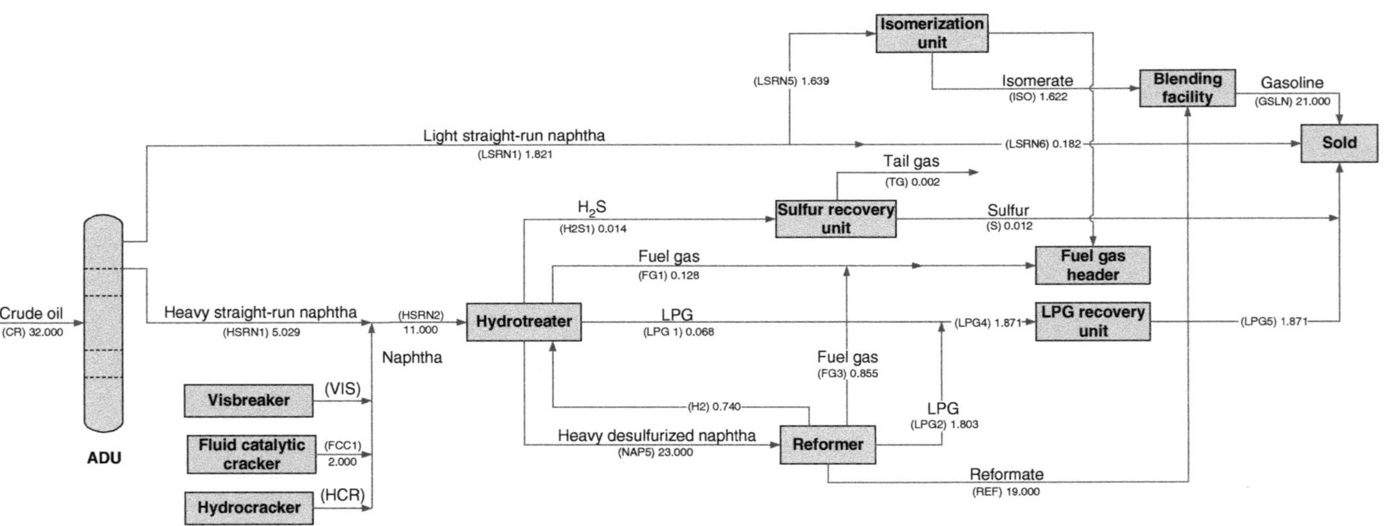

Figure 5.9 GDP result on optimal naphtha production configuration for heavy crude oil charge.

Table 5.13 GDP main results summary.

	Light crude oil	Heavy crude oil
CAPEX + OPEX + Raw material (mil RM)	2746	2747
CAPEX (mil RM)	791	791
OPEX + Raw material (mil RM)	1955	1951
Crude feed requirement (kg/d)	4.5E+7	3.2E+7
Raw material (mil RM)	42	30
OPEX (mil RM)	1912	1920
OPEX (mil RM/d)	5.7	5.8

5.6 Chapter Summary

The chapter presents optimization model formulations for petroleum refinery configuration design that are generic. The formulations give rise to two of such linear model variants, namely, MILP and GDP. We discuss the solution through computational results obtained from a numerical example to illustrate the model's features and applicability to practical industrial problems.

References

Dempster, M.A.H., Hicks Pedrón, N., Medova, E.A. et al. (2000). Planning logistics operations in the oil industry. *Journal of the Operational Research Society* 51 (11): 1271–1288.

Drud, A.S. (2021). CONOPT. Fairfax, VA: GAMS.

Edgar, T.F., Himmelblau, D.M., and Lasdon, L.S. (2001). *Optimization of Chemical Processes*, 2e, 326–327. New York, USA: McGraw-Hill.

Escudero, L.F., Quintana, F.J., and Salmerón, J. (1999). CORO, a modeling and an algorithmic framework for oil supply, transformation and distribution optimization under uncertainty1This work has been partially supported by the European Commission within the ESPRIT program HPC domaine, project HChLOUSO ES24897.1. *European Journal of Operational Research* 114 (3): 638–656.

Gary, J.H., Handwerk, G.E., and Kaiser, M.J. (2007). *Petroleum Refining: Technology and Economics*, 5e. New York: Marcel Dekker.

Gurobi Optimization (2020). Gurobi Optimizer Reference Manual. http://www.gurobi .com.

Hooker, J.N., Yan, H., Grossmann, I.E., and Raman, R. (1994). Logic cuts for processing networks with fixed charges. *Computers & Operations Research* 21 (3): 265–279.

Hydrocarbon Processing magazine (2006). HPI Construction Boxscore.

IBM. *CPLEX Performance Tuning for Mixed Integer Programs* (2019). Available from https://www.ibm.com/support/pages/cplex-performance-tuning-mixed-integer-programs (accessed 11 November 2021).

IBM (2020) *IBM ILOG CPLEX Optimization Studio V12.9.0* 2020. https://www.ibm.com/support/knowledgecenter/SSSA5P_12.9.0/ilog.odms.studio.help/Optimization_Studio/topics/COS_home.html (accessed 11 November 11 2021).

Karmarkar, N. (1984). A new polynomial-time algorithm for linear programming. *Combinatorica* 4 (4): 373–395.

Kovac, M., Movik, G., and Elliot, J.D. (2006). Upgrade refinery residuals into value-added products. *Hydrocarbon Processing* 86 (6): 57–62.

Malaysian Industrial Development Authority (MIDA). 2008. Petroleum and Petrochemicals Industry Data (2008) for Malaysia.

Maples, R.E. (2000). *Petroleum Refinery Process Economics*, 2nde, 264. Pennwell: Oklahoma.

Parkash, S. (2003). *Refining Processes Handbook*. Burlington: Gulf Professional Publishing.

Pongsakdi, A., Rangsunvigit, P., Siemanond, K., and Bagajewicz, M.J. (2006). Financial risk management in the planning of refinery operations. *International Journal of Production Economics* 103 (1): 64–86.

Raman, R. and Grossmann, I.E. (1991). Relation between MILP modelling and logical inference for chemical process synthesis. *Computers & Chemical Engineering* 15 (2): 73–84.

Raman, R. and Grossmann, I.E. (1992). Integration of logic and heuristic knowledge in MINLP optimization for process synthesis. *Computers & Chemical Engineering* 16 (3): 155–171.

Raman, R. and Grossmann, I.E. (1993). Symbolic integration of logic in mixed integer linear programming techniques for process synthesis. *Computers & Chemical Engineering* 17: 909–927.

Rardin, R.L. (1998). *Optimization in Operations Research*. New Jersey: Prentice-Hall.

Sahinidis, N.V., Grossmann, I.E., Fornari, R.E., and Chathrathi, M. (1989). Optimization model for long range planning in the chemical industry. *Computers & Chemical Engineering* 13 (9): 1049–1063.

Turkay, M. and Grossmann, I.E. (1996). Logic-based MINLP algorithms for the optimal synthesis of process networks. *Comput Chem Eng* 29: 959.

Yeomans, H. and Grossmann, I.E. (1999). A systematic modeling framework of superstructure optimization in process synthesis. *Computers & Chemical Engineering* 23: 709–731.

6

Solution Strategies

This chapter offers available solution strategies that are potentially tractable to address large-scale nonlinear nonconvex optimization problems. Most practical optimization problems entail many variables and constraints. Consequently, the solution time of such models increases rapidly with the problem size and poses convergence challenges. Furthermore, there may be no guarantee of global optimal solutions for nonconvex optimization problems (Floudas and Gounaris 2008). In this regard, solution strategies are available including convex relaxation, Lagrangean decomposition, and a few other global optimization techniques, which are surveyed in this chapter. More details are discussed in specialized texts on this subject (Hansen 1992; Horst and Tuy 1993; Horst et al. 1995; Floudas 2000).

6.1 Convex Relaxation

A common technique for relaxing nonconvexities (hence called convex relaxation) involves replacing them by using over- and under-estimating functions. A standard approach is by employing convex envelopes (McCormick 1976) for bilinear terms, which gives the tightest possible of such convex extension or relaxation of a feasible set (although their construction is not always guaranteed) (Al-Khayyal and Falk 1983; Floudas and Gounaris 2008; Gounaris et al. 2009). The literature on constructing convex relaxation is mainly focused on nonconvex nonlinear programming (LP) problems that involve factorable functions, which can be expressed as products of univariate functions (e.g. for bilinear functions as shown in Figure 6.1) or recursive sums (Bergamini et al. 2008). More recent work includes using piecewise linear or piecewise-affine relaxation schemes with numerous variants that differ in the statistics of the reformulated problem besides computational performance in terms of number of iterations and convergence speed (Wicaksono and Karimi 2008; Gounaris et al. 2009; Baltean and Misener 2018).

Model-Based Optimization for Petroleum Refinery Configuration Design, First Edition. Cheng Seong Khor.

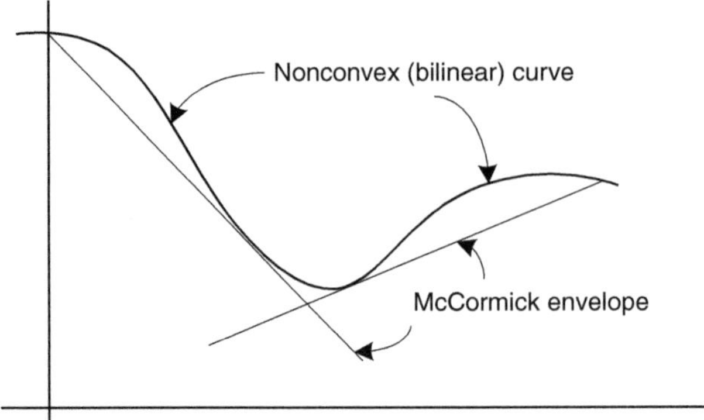

Figure 6.1 Convex envelopes for nonlinear bilinear terms. Source: Adapted from McCormick (1976).

6.2 Lagrangean Decomposition

Decomposition generally involves reformulating a large problem into several small ones by exploiting the problem structure. The motivation stems from that it takes less time to solve several small problems repeatedly than to do so on a large problem once. The Lagrangean decomposition method decomposes a problem into smaller subproblems by identifying what are called complicating variables. For a minimization problem, a Lagrangean subproblem solution gives a lower bound to the solution of an original problem. The bounds quality of Lagrangean decomposition depends on the variable bounds, scaling, and decomposition method (Karuppiah and Grossmann 2008; Yang et al. 2014). Figure 6.2 illustrates a conceptual sketch of a Lagrangean decomposition scheme (Daoutidis et al. 2019).

6.3 Global Optimization Techniques

Several global optimization techniques, tools, and procedures have been devised by incorporating fundamental principles from the strategies of convex relaxation and Lagrangean decomposition often within a branch-and-bound (Land and Doig 1960) framework in tandem with other techniques as surveyed next. This research area has received attention not only from that of process systems engineering (PSE)

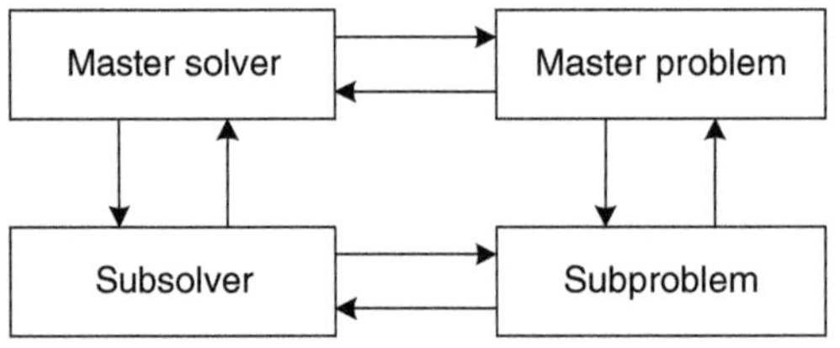

Figure 6.2 An illustration of a Lagrangean decomposition scheme. Source: Adapted from Daoutidis et al. (2019).

Table 6.1 Various global optimization techniques (with emphasis on MINLP problems).

Technique	Strategy	Problem type	Representative work
Branch and bound	Branching on both integer and continuous variables within branch-and-bound framework	Integer binary variables; nonconvex functions in continuous variables with continuous second-order derivatives	Adjiman et al. (2000)
Branch and reduce	Extends branch-and-bound framework with several tests to accelerate solution space reduction	Separable bilinear problems with valid convex underestimators	Ryoo and Sahinidis (1995)
Reformulation/ Spatial branch and bound	Automatically reformulates original nonlinear problems into those with special structure by introducing new constraints and variables solved using branch and bound with convex relaxation at each node	Convex or concave functions involving binary arithmetic	Smith and Pantelides (1999)
Hybrid branch and bound	Performs convexification, relaxation, and transformation to build convex underestimators	Nonlinear equality and inequality constraints	Zamora and Grossmann (1998b)
Extended cutting plane	Linearizes original nonlinear problems (without underestimators) and generates new constraints for increasingly tighter linearized problem	Pseudoconvex nonlinear problems (MINLP)	Westerlund et al. (1994, 1998)
Interval analysis	Branching on both integer and continuous variables; computes guaranteed ranges using successive domain partitioning and bounding	Twice-differentiable objective and once-differentiable constraints	Vaidyanathan and El-Halwagi (1996)

Source: Adapted from Floudas (2009).

particularly for applied problems (Floudas and Gounaris 2008; Misener et al. 2010) but also several communities as this field continues to be actively pursued (Misener and Floudas 2012; Boukouvala et al. 2016; Li et al. 2016; Lara et al. 2018). Table 6.1 presents a brief summary of the main methods proposed and the problem type or structure for which they are employed. Other references on this subject especially those relevant to PSE are available in the literature (Grossmann 1996; Tawarmalani and Sahinidis 2002, 2004).

6.3.1 Branch and Reduce

The branch-and-reduce technique adopts the branch-and-bound procedure in constructing valid convex underestimators to relax separable nonlinear bilinear problems. Branching is performed on continuous variables to devise underestimation constraints that generate lower bounds of the objective function values of the nonconvex nonlinear terms. Box reduction constraints are incorporated to tighten the bounds. To expedite solution space reduction, several tests for feasibility or optimality are proposed (Ryoo and Sahinidis 1995) (Figure 6.3).

6.3.2 Spatial Branch and Bound

The reformulation-cum-spatial branch-and-bound approach applies to nonlinear problems with convex or concave objective and constraints involving binary arithmetic operations (including combinations thereof). The algorithm first reformulates a nonlinear problem to involve only convex or concave terms (e.g. linear, linear fractional, bilinear, and simple exponentiation) by introducing new constraints and variables. A branch-and-bound procedure is then implemented to solve the reformulated problem to global optimality by exploiting its special structure, which allows constructing convex relaxation at every node of the tree. This approach can result in a large reformulated relaxed problem even for modest size problems (Smith and Pantelides 1996, 1999).

6.3.3 Hybrid Branch and Bound

The hybrid branch-and-bound approach is used in conjunction with other techniques, e.g. outer approximation (Duran and Grossmann 1986) to perform convexification, relaxation, and transformation to build convex underestimators for optimizing nonlinear optimization problems (e.g. mixed-integer nonlinear programming [MINLP]). Such underestimation schemes subsequently render solving the resulting convex model reformulation using outer approximation

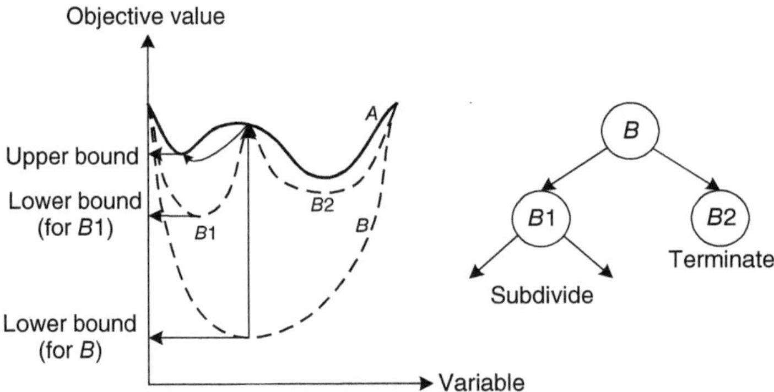

Figure 6.3 Illustration of the branch-and-reduce technique. Source: Adapted from Ryoo and Sahinidis (1995).

method (Zamora and Grossmann 1997, 1998a,b). Another variant of the hybrid branch-and-bound technique can be found in an algorithm called αBB to handle continuous twice-differentiable problems (NLP), which are linear and separable involving binary or integer variables (Androulakis et al. 1995; Adjiman et al. 2000). Other related variants of a hybrid branch-and-bound approach include branch and cut (Padberg and Rinaldi 1991), branch and price, branch and infer (Van Hentenryck et al. 1997), and combination of these techniques such as branch, cut, and price.

6.3.4 Interval Analysis

The interval analysis-based approach applies to problems with twice-differentiable objective function and once-differentiable constraints. Branching is performed on both integer and continuous variables to partition the domain and objective function bounds successively. The method involves computing guaranteed ranges for the functions. Unlike branch-and-bound-based algorithms, the problem solution bounds in a given domain are determined based on the objective function range (and not through optimization), which might render them to be loose, hence efficient fathoming (pruning) techniques are required to attain convergence (Moore 1979; Ratschek and Rokne 1984; Neumaier 1991).

6.3.5 Extended Cutting Plane

The extended cutting plane approach applies to pseudo- or quasi-convex nonlinear problems including MINLP. It mainly differs from other methods in that it involves a linearization procedure that generates new linear constraints without constructing underestimators at each of the iterations (as roughly illustrated in Figure 6.4) and adopts a modified convergence criterion (Westerlund et al. 1994, 1998). This approach generates a linear constraint (i.e. cut) by solving mixed-integer subproblems of linearized nonlinear objective or constraint functions at candidate solution points in each iteration.

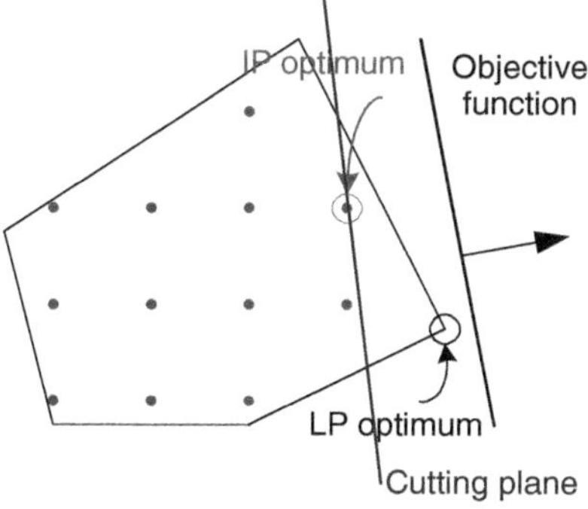

Figure 6.4 A general representation of cutting plane approach.

6.4 Advancements in Commercial Integer Optimization Solvers

This section focuses on advancements in commercial integer optimization solvers as exemplified by the CPLEX software package particularly for (but not limited to) MILP models applied to refinery problems. Background on the main underlying algorithmic method of branch and cut, which is based on the established optimization solution methods of branch-and-bound and cutting planes as introduced earlier in this chapter. We also cover heuristic-based algorithms, which include preprocessing and probing strategies as well as the more advanced methods of local or neighborhood search for polishing solutions toward enhanced use in practical settings. Emphasis is given to both theory and implementation of the methods available. Other considerations are offered on techniques such as parallelization, solution pools, and tuning tools to improve computational performance.

6.4.1 Overview

Numerous technical- and business-oriented applications can be posed as mathematical programming problems that can be handled by commercial optimization solvers such as CPLEX (IBM 2020), Gurobi (Gurobi Optimization 2020), or KNITRO (Byrd et al. 2006). The problems can be formulated as models that include LP, MILP, quadratic programming (QP), mixed-integer quadratic programming (MIQP), quadratically constrained programming, and mixed-integer quadratically constrained programming. Such solvers are also used in tandem with other appropriate optimization solvers to handle other mainly nonlinear problems such as MINLP models or in general, mixed-integer programs (MIPs) (Jünger et al. 2010).

6.4.2 Computational Performance of Commercial Integer Optimization Solvers

The actual computational performance of a commercial optimizer (or optimization package) such as CPLEX results from a combination of improvement in several aspects. They include LP solvers with capability and features including preprocessing, algebra for sparse systems, solution methods (primal or dual simplex and barrier), and techniques to overcome degeneracy and numerical difficulties (Rardin 1998). Equally important is the use of cutting planes as valid inequalities in solving problems that bridge the gap from theory to practice (Williams 1999). Further improvement involves applying heuristics including node heuristics (e.g. local branching, guided dives) and relaxation-induced neighborhood search (RINS), invoking evolutionary algorithms for solution polishing; and implementing parallelization for efficient computations (Danna 2008).

6.4.3 A Commercial Success Story: CPLEX Integer Optimization Solver

CPLEX is a state-of-the-art commercial integer optimization solver currently marketed by IBM. It represents an early commercial success story of an optimization

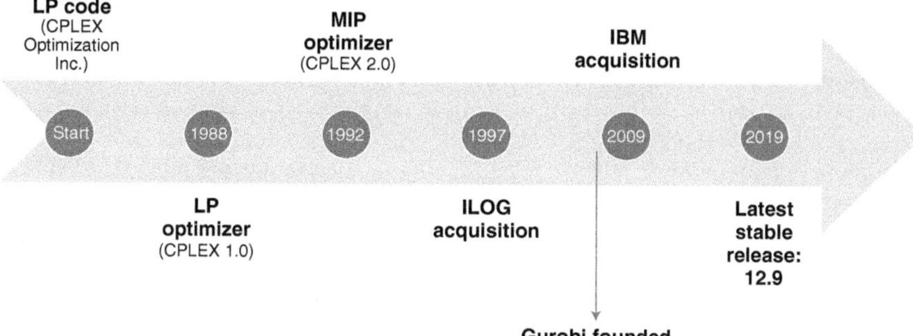

Figure 6.5 Historical background of IBM ILOG CPLEX integer optimization solver.

Table 6.2 Software release history of IBM ILOG CPLEX integer optimization solver.

Year	Activity/accomplishment
1988	Develops LP solver (CPLEX 1.0)
1992	Offers simple branch and bound with limited cuts (CPLEX 2.0)
1998	Incorporates simple heuristic; provides faster dual simplex (CPLEX 6.0)
1999	Introduces five node heuristics and six cutting plane types (CPLEX 6.5)
2000	Caters for semi-continuous and semi-integer variables; stipulates dual simplex as default LP solution method; introduces preprocessing; improved cuts (CPLEX 7.0)
2002	Introduces new LP method of sifting, concurrent optimization, new QP capabilities, and 9 cutting plane types (CPLEX 8.0)
2003	Introduces quadratic constraint programming (QCP) and relaxation induced neighborhood search (RINS) (CPLEX 9.0)
2006	Improved MIQP, changes in MIP start, feasible relaxation; introduces indicators and solution polishing features
2007	CPLEX 11.0 incorporates solution pool, tuning pool, and parallel mode
2010	Offers faster MILP solution; introduces multicommodity flow cuts; enhanced heuristics and dynamic search (in CPLEX 12.2)
2017	Enables faster MILP solution; enhanced CP Optimizer; new callback framework (in CPLEX 12.8)
2019	Includes handling of multiobjective problems; provides modeling assistance (in CPLEX 12.9)

Source: Khor (2021). Reproduced with permission of IntechOpen.

package with various acquisitions and a spin-off solver called Gurobi which is now a success story of its own. Figure 6.5 presents brief historical facts of CPLEX, while Table 6.2 summarizes the software release history.

6.4.4 Solution Methods and Algorithms

6.4.4.1 Integer Optimization Algorithms

A suite of algorithms is available in various integer optimization solvers to exploit the underlying problem structure of an application toward achieving efficiency and

Table 6.3 Algorithms available in IBM ILOG CPLEX integer optimization solver.

Solver/optimizer	Algorithm	Model type	Remark
Simplex	Primal, dual, network	LP, QP	Reoptimization with simplex algorithms is faster when starting from a previous basis
Barrier	Interior-point	LP, QP, QCP	Explore multiple threads presence
			Barrier optimizer cannot start from advanced basis—limited application in B&B for MILP
Mixed-integer optimizers	Branch and cut, dynamic search	MILP, MIQP, MIQCP	IBM proprietary/trade secret methodology to solve MIP (some user callbacks cannot be used)

Source: Khor (2021). Reproduced with permission of IntechOpen.

accuracy. Table 6.3 summarizes the typical main algorithms employed by CPLEX according to the problem type identified together with remarks on the enhancement provided to increase computational performance (Rothberg 2003b).

6.4.4.2 Branch and Bound

A general structure of MIP is given by:

$$\min \quad c^T x \tag{6.1}$$

$$\text{s.t.} \quad Ax = b \tag{6.2}$$

$$l \leq x \leq u \tag{6.3}$$

$$\text{some } x \text{ are integers} \tag{6.4}$$

Branch and bound is a base algorithm to solve MIP which uses LP as a subroutine (Land and Doig 1960). The key strategies of a branch-and-bound procedure involve splitting (i.e. branching) the solution space into disjoint subspaces, bounding the objective function values for all solutions in the subspaces, and pruning or fathoming nodes of branches that cannot yield better solutions. Although it is provably exponential in time, tricks are available to accelerate its search which mostly applies to a subset of models with a suite of algorithms available.

The branching strategies are performed on the integer variables and comprise two main steps: (1) Choose an integer variable as a branching variable x_j, (2) Split the problem into two submodels: $x_j \leq i$ or $x_j \geq i+1$ where for the special case of binary variables, the problem becomes $x_j = 0$ or $x_j = 1$.

The bounding problem is given by the continuous (LP) relaxation to determine a lower bound z_{IP}^L on the objective function value of the original MIP problem can be described as follows: minimize $c^T x$ $\left(= z_{IP}^L\right)$ subject to $Ax = b$, $l \leq x \leq u$ (simple bounds), and some x_j are integers. The continuous relaxation problem gives

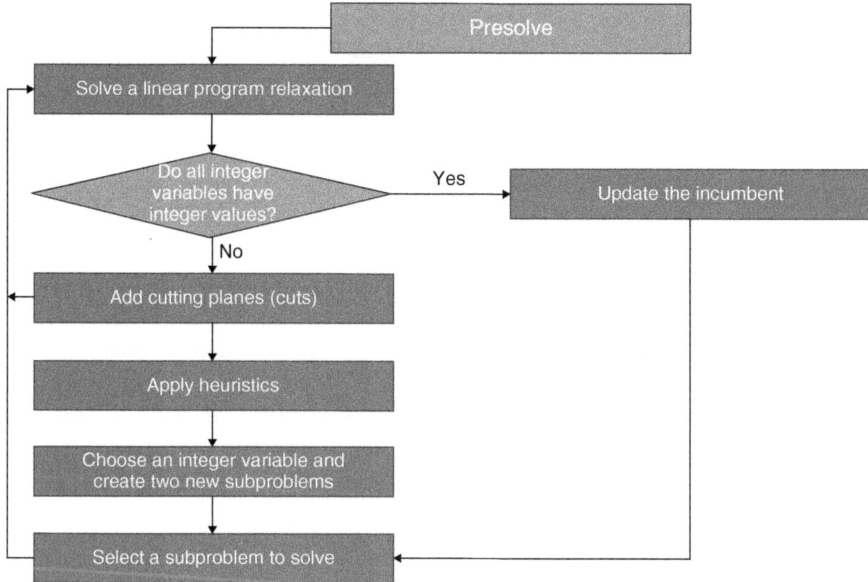

Figure 6.6 Branch and cut algorithm. Source: Adapted from CPLEX.

solution of an optimal objective value of z_{IP}^{L}, which is a lower bound on the objective function value of the original MIP problem by relaxing the integrality restriction. There are two useful properties of continuous relaxation: (1) if its solution satisfies integrality restrictions, there is no need to further explore the subspace; (2) it offers natural branching candidates as the integer variables with fractional values in a relaxation solution.

Key steps in the branch-and-bound procedure are described in Figure 6.6. The node selection in step 1 involves a tradeoff between achieving feasibility and optimality. The options available for node selection include depth first, breadth first, best first, limited discrepancy, and best estimate. When exploring nodes deep in a search tree, one is more likely to find integer feasible solutions and explore nodes that would be pruned by later feasible solutions. The method called plunging (as combined with those aforementioned) always chooses a child node of a previously explored node.

In step 2, the node relaxation step is ideally suited to dual simplex method. It involves only a small change from the parent relaxation solution (at the root node) and gives a new bound on the branching variable while maintaining the dual feasibility of the previous basis. Thus, the solution is likely to be close to the previous basis. Typically, a few dual simplex iterations are sufficient to restore optimality, and the cost per node is quite small. The subsequent step 3 entails generating cutting planes as needed to obtain a continuous (LP) relaxation solution.

Step 4 involves variables fixing using reduced cost. If the following condition as given by Eq. (6.5) holds at a branch-and-bound node:

$$z_{LP} + |D_j| \geq z^* \tag{6.5}$$

where z_{LP} = objective value of LP relaxation solution at the root node, z^* = objective value of an incumbent (i.e. best-known integer feasible solution), and D_j = reduced cost (marginal cost of releasing a variable from its bound), then we apply the strategy of fixing x_j to its current value in this subtree of the search. The goal here as described by step 5 is to obtain integer feasible solutions which are like the relaxation solution.

Selecting an appropriate branching variable can significantly affect the search tree size, which is emphasized in the subsequent step 6. In this regard, the guiding principles are to make important decisions early (as modeled by the integral branching variables) by being aware of the impact of both branching directions. To illustrate by using a factory building problem, such a decision involves whether to build a factory first while the decision on the number of lines to be placed in the factory can be made later. In general, we can predict the impact of a branch by considering variables that are furthest from their bounds which indicate maximum infeasibility. Thus, the impact for each branching candidate can be measured to allow for strong branching to be performed, e.g. by using historical information such as pseudo-costs.

Finally, in step 7, the main idea in propagating implications logically is to fix the binary variables to possible values during tree exploration and determine the binary variable values. Bound strengthening is used to tighten variable bounds.

Practical considerations render implementing branch and bound to be unsuitable for large-scale problems chiefly because the number of iterations grows exponentially with number of variables. Therefore, in practice, a commercial business intelligence solver such as CPLEX uses a branch-and-cut procedure as a modification which applies model reformulation by using presolve strategies and adding cutting planes (or cuts) as shown in Figure 6.4 with possible enhancement in practice around the root node computations (Lima and Grossmann 2017).

6.4.4.3 Presolve and Cutting Planes

The original MIP formulation can be improved by tightening it with fewer constraints and variables thus entailing less data handling requirement (yet with the same solution quality). A tighter formulation also leads to a smaller difference between the space of the feasible continuous and feasible integer solutions, hence relying less on branching to refine the continuous relaxation computation. Two techniques are used: (1) presolve which combines preprocessing and probing strategies (Savelsbergh 1994; Wolsey 1998); and (2) cutting planes (Nemhauser and Wolsey 1988).

Presolve generates a new tighter improved model without a size increase that is independent of the relaxation solution. Preprocessing aims to identify feasibility and redundancy while improving bounds (e.g. through rounding), while probing improves coefficients by fixing the binary variable values while checking for their logical implications. In both cases, we achieve a tighter model reformulation using similar steps of adding or replacing constraints that maintain the same integer solutions but with fewer continuous relaxation solutions. Adding a single constraint can produce an exponential number of tighter constraints. Such tighter constraints dominate the existing constraints without creating a larger problem. Note that reformulation solution is different from that of relaxation.

In contrast, we add a cutting plane (or valid inequality) to an existing model (typically the presolve-reformulated model) to remove a relaxation solution – this feature constitutes an important difference between the two techniques. Therefore, cutting planes introduce tighter constraints that cut off a particular relaxation solution and in so doing, achieves focused growth in model size.

In summary, presolve is vital in solving MIP as there is significant scope to improve most model formulations through reducing problem sizes (by more than five times is not uncommon) or runtimes (similarly by up to 10 times). On the other hand, cutting planes are available in numerous varieties with many valid types applicable for a particular model. Thus, we need to identify relevant ones which serve to cut off appealing relaxation solutions. There is a need to strike a balance in terms of how many cuts to generate for a relaxation solution. Since we need to cut off relaxation solution only once, and it is expensive to resolve in obtaining a new relaxation solution for each cut added, we conduct multiple rounds of cutting plane generation while limiting the number of cuts per round in view of the increased model size (Rothberg 2003c).

6.4.4.4 Heuristics

Heuristics for solving MIP aims to produce good and possibly feasible solutions quickly without relying on branching in satisfying user demands for a problem. Thus, heuristics avoid exploring unproductive subtrees (in a branch-and-cut scheme) while exploring parts of tree that a solver typically will not. In doing so, heuristics help to prove optimality explicitly by pruning nodes more efficiently as well as implicitly by giving integer solutions (Rothberg 2003a).

Heuristics can be classified into two classes as available in a solver like CPLEX: (1) plunging (diving) heuristics, and (2) local improvement heuristics which explore interesting neighborhoods around potential solutions using search strategies such as local branching, RINS, guided dives, and evolutionary algorithms for solution polishing. Plunging heuristics maintains linear feasibility in trying to achieve integer feasibility, while local improvement heuristics operate conversely (Rothberg 2007). A typical strategy for heuristics applied at the root node involves the sequence shown in Figure 6.4.

Some considerations in applying plunging heuristics include tradeoffs of how many variables to fix per computation round and in what order. While it is computationally inexpensive to fix all variables rather than a few variables, LP relaxation solutions in the latter (not needed in the former) can guide later choices (e.g. on variable values and reduced costs). Variations in variable fixing order can be useful for diversification. On the other hand, a high-level structure of local improvement heuristics involves choosing integer values for all the integer variables, which produces linear infeasibility; iterating over the integer variables; and applying infeasibility metrics (Rothberg 2003a).

The effectiveness of heuristics is evidenced in that feasible solutions are found for most models before branch and bound is performed. Approximately 10 % improvement in computational time to proven optimality has been reported (Rothberg 2003a). Furthermore, heuristics often get solutions not obtained by branching.

6.4.4.5 Combined Local Search and Heuristics

A combination of local search and heuristics offers a powerful optimization framework to solve difficult MIP or combinatorial optimization problems. Examples of local search methods include simulated annealing, tabu search, and genetic algorithms. Local search methods consist of the key strategies of neighborhood (i.e. considers a set of solutions in the vicinity of current solution); intensification (i.e. temporary focus on part of solution space); and diversification (i.e. mechanism to change focus occasionally). In applying local search to MIP, generally neighborhoods are based on the problem structure, e.g. nodes and edges in graphs with no high-level structural information available in arbitrary MIP models (Rothberg 2003a). A question that arises is how we can generate and explore an interesting neighborhood given an incumbent solution. In this regard, two methods are available, namely, local branching (Fischetti and Lodi 2003) and RINS (Danna et al. 2005).

6.4.4.6 Parallelization

Parallelization is available in an integer optimization solver such as CPLEX, which encompasses the MIP solution engine, barrier algorithm, and concurrent optimization techniques for solving LP and QP problems. In the instance of CPLEX, parallelization involves launching several optimizers to solve the same problem – the process stops when the first solver reaches a solution. Within a branch-and-bound scheme, parallelization involves solution of the root node and nodes as well as strong branching in parallel (Lima 2010).

6.4.4.7 Solution Pools

The motivation to consider solution pools lies in the value of having more than one solution due to inaccurate data, approximations in model formulations, or inability of a model to capture the full essence of a problem. Thus, solution pools aim to generate and keep multiple solutions by using various options and tools that involve collecting solutions within a given percentage of optimal solution or those with diverse solutions and properties. However, the difficulty is noted in implementing solution pools with the strategy of rolling horizon decompositions (Rothberg 2007).

6.4.4.8 Tuning Tools

As MIP solvers have multiple algorithm parameters which dictate their performance, the objective of the tuning tool is to identify solver parameters that improve the performance for a given problem set. While default parameter values of MIP solvers are defined to work well for a large collection of problems, there is no such guarantee generally for a specific user problem (Bixby et al. 2000).

6.4.5 Application Examples

This section presents three snippets of examples of applying commercial integer optimization solvers to implement and improve or enhance business intelligence applications. The model formulations for the examples are implemented on the GAMS modeling platform (and available in GAMS Model Library) from which the CPLEX solver is accessed.

6.4.5.1 Example 1: Energy Optimization

The first example presents a practical application of CPLEX as a standard solver for an energy business portfolio optimization problem for an electric utility company (Rebennack et al. 2010). For such electricity distribution public service, the problem involves determining the amount to produce internally (i.e. in one's own power plant) and that to purchase externally (i.e. from the spot market or load following contracts). The problem formulation leads to a medium-to-large scale MILP model with size and computational statistics as described in Table 6.4. To accelerate solution convergence, several computational options are invoked including priority branching within a branch-and-bound procedure and multiple processing through parallelization (i.e. techniques introduced in the foregoing section). The computational results and implications as discussed in the cited reference demonstrate the applicability of the solver as an effective tool for one-day ahead planning within a real-world electricity market in Germany.

6.4.5.2 Example 2: Financial Optimization

The second example involves financial optimization of risk management with commercial implications (Dahl et al. 1993). The problem is amenable to be posed as an integer optimization model to capture an extensive set of rules and regulations that governs the delivery and settlement of mortgage-backed securities. The availability of reliable, robust, and efficient commercial integer optimization solvers alongside computing technology developments has facilitated the deployment and validation of such models with the computational statistics summarized in Table 6.5. The advancement achieved has led to optimization models including (if not particularly) integer programs to become essential omnipresent tools in current financial operations, which is comparable to the application of operations research and management science models in the domains of manufacturing, transportation, and logistics.

6.4.5.3 Example 3: Manufacturing Optimization

The third example concerns production planning for a manufacturing facility (Pochet and Wolsey 2006). The application can be formulated as a standard integer

Table 6.4 Model size and computational statistics for Use Case 1.

Computing platform	GAMS 24.2.3 on laptop with Intel Core i7-8550U 1.80 (up to 1.99) GHz, 8 GB of RAM
No. of continuous variables	1260
No. of discrete variables	773
No. of constraints	2178
No. of iterations	2,799,216
CPU time	408.234 second
Objective function value	EUR266,793 (for optimality gap = 0%)

Source: Khor (2021). Reproduced with permission of IntechOpen.

Table 6.5 Model size and computational statistics for Use Case 2.

Computing platform	GAMS 24.2.3 on laptop with Intel Core i7-8550U 1.80 (up to 1.99) GHz, 8 GB of RAM
No. of continuous variables	255
No. of discrete variables	199
No. of constraints	487
No. of iterations	238
CPU time	0.157 second
Objective function value	36.96 (for optimality gap = 1×10^{-4}%)

Source: Khor (2021). Reproduced with permission of IntechOpen.

Table 6.6 Model size and computational statistics for Use Case 3.

Computing platform	GAMS 24.2.3 on laptop with Intel Core i7-8550U 1.80 (up to 1.99) GHz, 8 GB of RAM
No. of continuous variables	81
No. of discrete variables	8
No. of constraints	73
No. of iterations	15
CPU time	0.016 second
Objective function value	36.96 (for optimality gap = 0^4%)

Source: Khor (2021). Reproduced with permission of IntechOpen.

optimization model of an uncapacitated lot-sizing problem. The objective function seeks to minimize production cost in meeting market demand constraints with cost components on production, stocking, and machine setups. Table 6.6 gives the model size and computational statistics for the largest problem instance solved for this example.

6.4.5.4 Concluding Remarks

Performance variability across commercial integer optimization solvers applied to business intelligence applications (such as that for the examples in Section 6.4.5) occurs due to opportunistic parallelization, use of heuristics particularly by invoking polishing option (which involves random seed), or simply numerical reasons. Variability may be observed in computational time, performance in terms of number of nodes and iterations, or solution quality. A main limitation of the applicability of integer optimization solvers typically pertains to the number of integer variables that can be handled within an acceptable computational load or solution time. Therefore, it is worthwhile for future research in this area to consider further improvement in the mentioned areas (Bixby et al. 2000; Bixby and Rothberg 2007) toward achieving acceptable performance levels that are requisite and crucial for business intelligence applications.

6.5 Chapter Summary

This chapter aims to give an overview on the standard techniques for handling nonconvex nonlinear process engineering problems as briefly summarized in Table 6.1. Such problems are commonly encountered in the PSE domain, which leads to the contribution by the community to global optimization approach. Several other techniques have emerged over the years particularly variants based on the branch-and-bound principle such as branch and reduce (Sahinidis 2009). This area remains active, fertile, and important for both academic research and practical applications.

References

Adjiman, C.S., Androulakis, I.P., and Floudas, C.A. (2000). Global optimization of mixed-integer nonlinear problems. *AIChE Journal* 46 (9): 1769–1797.

Al-Khayyal, F.A. and Falk, J.E. (1983). Jointly constrained biconvex programming. *Mathematics of Operations Research* 8 (2): 273–286.

Androulakis, I.P., Maranas, C.D., and Floudas, C.A. (1995). αBB: a global optimization method for general constrained nonconvex problems. *Journal of Global Optimization* 7 (4): 337–363.

Baltean, L. and Misener, R. (2018). Piecewise parametric structure in the pooling problem – from sparse, strongly-polynomial solutions to NP-hardness. *Journal of Global Optimization* 71: 655–690.

Bergamini, M.L., Grossmann, I., Scenna, N., and Aguirre, P. (2008). An improved piecewise outer-approximation algorithm for the global optimization of MINLP models involving concave and bilinear terms. *Computers & Chemical Engineering* 32 (3): 477–493.

Bixby, R. and Rothberg, E. (2007). *Progress in computational mixed integer programming—a look back from the other side of the tipping point. Annals of Operations Research* 149 (1): 37–41.

Bixby, R., Fenelon, M., Gu, Z. et al. (2000). MIP: theory and practice – closing the gap. In: *System Modelling and Optimization: Methods, Theory, and Applications* (ed. M.A. Boston), 19–49. Kluwer Academic Publishers.

Boukouvala, F., Misener, R., and Floudas, C.A. (2016). Global optimization advances in mixed-integer nonlinear programming, MINLP, and constrained derivative-free optimization, CDFO. *European Journal of Operational Research* 252 (3): 701–727.

Byrd, R.H., Nocedal, J., and Waltz, R.A. (2006). KNITRO: an integrated package for nonlinear optimization. In: *Large-Scale Nonlinear Optimization* (ed. G.d. Pillo and M. Roma), 35–59. Springer.

Dahl, H., Meeraus, A., and Zenios, S.A. (1993). Some financial optimization models: I Risk management. In: *Financial Optimization* (ed. S.A. Zenios), 3–36. Cambridge: Cambridge University Press.

Danna, E. (2008). Performance variability in mixed integer programming. *MIP 2008 Workshop*. New York City.

Danna, E., Rothberg, E., and Pape, C.L. (2005). Exploring relaxation induced neighborhoods to improve MIP solutions. *Mathematical Programming* 102 (1): 71–90.

Daoutidis, P., Tang, W., and Allman, A. (2019). Decomposition of control and optimization problems by network structure: concepts, methods, and inspirations from biology. *AIChE Journal* 65 (10): e16708.

Duran, M.A. and Grossmann, I.E. (1986). Simultaneous optimization and heat integration of chemical processes. *AIChE Journal* 32: 123.

Fischetti, M. and Lodi, A. (2003). Local branching. *Mathematical Programming* 98 (1): 23–47.

Floudas, C.A. (2000). *Deterministic Global Optimization: Theory, Algorithms and Applications*. Kluwer.

Floudas, C.A. (2009). Mixed integer nonlinear programming. In: *Encyclopedia of Optimization* (ed. C.A. Floudas and P.M. Pardalos), 405–414. Boston, MA: Springer US.

Floudas, C.A. and Gounaris, C.E. (2008). A review of recent advances in global optimization. *Journal of Global Optimization* 45 (1): 3.

Gounaris, C.E., Misener, R., and Floudas, C.A. (2009). Computational comparison of piecewise-linear relaxations for pooling problems. *Industrial & Engineering Chemistry Research* 48 (12): 5742–5766.

Grossmann, I.E. (1996). *Global Optimization in Engineering Design*. Boston, MA: Springer US.

Gurobi Optimization (2020). Gurobi optimizer reference manual. www.gurobi.com (accessed 11 November 2021).

Hansen, E. (1992). *Global Optimization Using Interval Analysis*. Marcel Dekker.

Horst, R. and Tuy, H. (1993). *Global Optimization*. Springer.

Horst, H., Pardalos, P.M., and Thoai, V. (1995). *Introduction to Global Optimization*. Kluwer.

IBM (2020). IBM ILOG CPLEX optimization studio V12.9.0. https://www.ibm.com/support/knowledgecenter/SSSA5P_12.9.0/ilog.odms.studio.help/Optimization_Studio/topics/COS_home.html.

Jünger, M., Liebling, T.M., Naddef, D. et al. (ed.) (2010). *50 Years of Integer Programming 1958–2008: From the Early Years to the State-of-the-Art*. Berlin Heidelberg: Springer-Verlag.

Karuppiah, R. and Grossmann, I.E. (2008). A Lagrangean based branch-and-cut algorithm for global optimization of nonconvex mixed-integer nonlinear programs with decomposable structures. *Journal of Global Optimization* 41 (2): 163–186.

Khor, C.S. (2021). Recent advancements in commercial integer optimization solvers for business intelligence applications. In: *E-Business - Higher Education and Intelligence Applications*. IntechOpen https://doi.org/10.5772/intechopen.93416.

Land, A.H. and Doig, A.G. (1960). An automatic method of solving discrete programming problems. *Econometrica* 28 (3): 497–520.

Lara, C.L., Trespalacios, F., and Grossmann, I.E. (2018). Global optimization algorithm for capacitated multi-facility continuous location-allocation problems. *Journal of Global Optimization* 71 (4): 871–889.

Li, J., Xiao, X., Boukouvala, F. et al. (2016). Data-driven mathematical modeling and global optimization framework for entire petrochemical planning operations. *AIChE Journal* 62 (9): 3020–3040.

Lima, R. (2010). IBM ILOG CPLEX: what is inside of the box? *Enterprise-Wide Optimization (EWO) Seminar*, Carnegie Mellon University, PA.

Lima, R.M. and Grossmann, I.E. (2017). On the solution of nonconvex cardinality Boolean quadratic programming problems: a computational study. *Computational Optimization and Applications* 66 (1): 1–37.

McCormick, G.P. (1976). Computability of global solutions to factorable nonconvex programs. 1. Convex underestimating problems. *Mathematical Programming* 10 (2): 147–175.

Misener, R. and Floudas, C.A. (2012). Global optimization of mixed-integer quadratically-constrained quadratic programs (MIQCQP) through piecewise-linear and edge-concave relaxations. *Mathematical Programming* 136 (1): 155–182.

Misener, R., Gounaris, C.E., and Floudas, C.A. (2010). Mathematical modeling and global optimization of large-scale extended pooling problems with the (EPA) complex emissions constraints. *Computers & Chemical Engineering* 34 (9): 1432–1456.

Moore, R.E. (1979). Applications to mathematical programming (optimization). In: *Methods and Applications of Interval Analysis*, 87–92. Society for Industrial and Applied Mathematics.

Nemhauser, G.L. and Wolsey, L.A. (1988). *Integer and Combinatorial Optimization*. Wiley-Interscience.

Neumaier, A. (1991). The solution of square linear systems of equations. In: *Interval Methods for Systems of Equations* (ed. A. Neumaier), 116–169. Cambridge: Cambridge University Press.

Padberg, M. and Rinaldi, G. (1991). A branch-and-cut algorithm for the resolution of large-scale symmetric traveling salesman problems. *SIAM Review* 33 (1): 60–100.

Pochet, Y. and Wolsey, L.A. (2006). *Production Planning by Mixed Integer Programming, Springer Series in Operations Research and Financial Engineering*. New York: Springer-Verlag.

Rardin, R.L. (1998). *Optimization in Operations Research*. New Jersey: Prentice-Hall.

Ratschek, H. and Rokne, J. (1984). *Computer Methods for the Range of Functions*. Chichester: Ellis Horwood.

Rebennack, S., Kallrath, J., and Pardalos, P.M. (2010). Energy portfolio optimization for electric utilities: case study for Germany. In: *Energy, Natural Resources and Environmental Economics* (ed. E. Bjørndal, M. Bjørndal, P.M. Pardalos, and M. Rönnqvist), 221–246. Berlin, Heidelberg: Springer Berlin Heidelberg.

Rothberg, E. (2003a). The CPLEX library: MIP heuristics. *4th Max-Planck Advanced Course on the Foundations of Computer Science (ADFOCS 2003)*. Saarbrücken, Germany: Max-Planck-Institut für Informatik.

Rothberg, E. (2003b). The CPLEX library: mixed integer programming. *4th Max-Planck Advanced Course on the Foundations of Computer Science (ADFOCS 2003)*. Saarbrücken, Germany: Max-Planck-Institut für Informatik.

Rothberg, E. (2003c). The CPLEX library: presolve and cutting planes. *4th Max-Planck Advanced Course on the Foundations of Computer Science (ADFOCS 2003).* Saarbrücken, Germany: Max-Planck-Institut für Informatik.

Rothberg, E. (2007). An evolutionary algorithm for polishing mixed integer programming solutions. *INFORMS Journal on Computing* 19 (4): 534–541.

Ryoo, H.S. and Sahinidis, N.V. (1995). Global optimization of nonconvex NLPs and MINLPs with applications in process design. *Computers & Chemical Engineering* 19 (5): 551–566.

Sahinidis, N.V. (2009). Global optimization. *Optimization Methods and Software* 24 (4): 479–482.

Savelsbergh, M.W.P. (1994). Preprocessing and probing techniques for mixed integer programming problems. *ORSA Journal on Computing* 6 (4): 445–454.

Smith, E.M.B. and Pantelides, C.C. (1996). Global optimisation of general process models. In: *Global Optimization in Engineering Design* (ed. I.E. Grossmann), 355–386. Boston, MA: Springer US.

Smith, E.M.B. and Pantelides, C.C. (1999). A symbolic reformulation/spatial branch-and-bound algorithm for the global optimisation of nonconvex MINLPs. *Computers & Chemical Engineering* 23 (4): 457–478.

Tawarmalani, M. and Sahinidis, N.V. (2002). *Convexification and Global Optimization in Continuous and Mixed-Integer Nonlinear Programming: Theory, Algorithms, Software, and Applications, Nonconvex Optimization and Its Applications*, vol. 65. Dordrecht: Kluwer Academic Publishers.

Tawarmalani, M. and Sahinidis, N.V. (2004). Global optimization of mixed-integer nonlinear programs: a theoretical and computational study. *Mathematical Programming* 99 (3): 563–591.

Vaidyanathan, R. and El-Halwagi, M. (1996). Global optimization of nonconvex MINLP's by interval analysis. In: *Global Optimization in Engineering Design* (ed. I.E. Grossmann), 175–193. Boston, MA: Springer US.

Van Hentenryck, P., Michel, L., and Deville, Y. (1997). *Numerica: A Modeling Language for Global Optimization.* Cambridge, MA: MIT Press.

Westerlund, T., Pettersson, F., and Grossmann, I.E. (1994). Optimization of pump configurations as a MINLP problem. *Computers & Chemical Engineering* 18 (9): 845–858.

Westerlund, T., Skrifvars, H., Harjunkoski, I., and Pörn, R. (1998). An extended cutting plane method for a class of non-convex MINLP problems. *Computers & Chemical Engineering* 22 (3): 357–365.

Wicaksono, D.S. and Karimi, I.A. (2008). Piecewise MILP under- and overestimators for global optimization of bilinear programs. *AIChE Journal* 54 (4): 991–1008.

Williams, H.P. (1999). *Model Building in Mathematical Programming*, 4e. Chichester, West Sussex, England: Wiley.

Wolsey, L.A. (1998). *Integer Programming*, Wiley-Interscience Series in Discrete Mathematics and Optimization. Chichester: Wiley.

Yang, L.L., Salcedo-Diaz, R., and Grossmann, I.E. (2014). Water network optimization with wastewater regeneration models. *Industrial & Engineering Chemistry Research* 53 (45): 17680–17695.

Zamora, J.M. and Grossmann, I.E. (1997). A comprehensive global optimization approach for the synthesis of heat exchanger networks with no stream splits. *Computers & Chemical Engineering* 21: S65–S70.

Zamora, J.M. and Grossmann, I.E. (1998a). Continuous global optimization of structured process systems models. *Computers & Chemical Engineering* 22 (12): 1749–1770.

Zamora, J.M. and Grossmann, I.E. (1998b). A global MINLP optimization algorithm for the synthesis of heat exchanger networks with no stream splits. *Computers & Chemical Engineering* 22 (3): 367–384.

7

Industrial Case Studies with Business-Centric Techno-Commercial Considerations

This chapter presents three case studies on refinery design configuration with business techno-commercial considerations.

7.1 Industrial Case Study 1: Refinery Configuration for Heavy Oil Processing

The need for heavy oil processing has increased in recent years worldwide, backed by higher demands for petroleum products in the face of declining light crude oil resources. The situation encourages refineries to focus more on maximizing the production of high-value outputs from these lower-value heavier feedstock. This example purports to assess heavy oil processing potential in the refining industry through model-based economic evaluation. We formulate a refinery model suitable for preliminary investment decision-making which considers various cost elements for several conventional commercial heavy oil processing technologies. The formulated model is applied to a case study on the worldwide potential for heavy oil processing. This example demonstrates the application of a model-based approach to perform or assist with investment assessment.

7.1.1 Background

Heavy crude oil upgrading has gained the interest of refineries as demand for petroleum products increased in the face of declining lighter crude oil resources. In today's market, there are abundant heavier crude oils in the market as compared to conventional lighter ones. However, multiple competing technologies exist with a wide range of product yields and energy (or utility) requirements to refine these heavy oil resources (Speight 2011a; Frecon et al. 2019).

Heavy crude oils contain high fractions of residue and are generally classified by the density measure of API gravity of less than 20. The residue requires additional upgrading processes to break the complex molecular structure in obtaining valuable products. Residue upgrading processes include several thermal and catalytic processes, which can be categorized as carbon rejection or hydrogen addition. Examples of carbon rejection processes are delayed coking (DCK), visbreaking (VB),

Model-Based Optimization for Petroleum Refinery Configuration Design, First Edition. Cheng Seong Khor.

fluid coking (FCK), and solvent deasphalting, while hydrogen addition technologies include fixed-bed hydroprocessing (e.g. Hyvahl F [HF]) and ebullated-bed hydroprocessing (e.g. LC Fining [LCF]) (Ancheyta and Speight 2007).

Integrating these technologies into refinery systems requires a systems approach-based economic evaluation instead of relying on, for example, monovariable decision-making such as solely based on attaining the highest product yields. The increasing demand for high-value petroleum products and declining for that of bottom distillate products encourage refineries to give more focus on maximizing the yields on heavier crudes. In addition to that, the price for heavy crude oils is generally lower than for lighter crude oils. Installing and operating heavy oil upgrading technologies enable refineries to buy cheaper feedstock and still produce high-value marketable products (Elshout et al. 2018).

There are available refinery optimization models of various complexities in terms of time and space scales, which gives rise to different computational requirements based on the purpose and activity. For high-level decision-making, linear programming (LP) models are suitable when only preliminary results are needed (Hartmann 1999; Zhang et al. 2001; Leiras et al. 2013; Albahri et al. 2019a; Khor 2019b). Nonlinear and/or mixed-integer models have been proposed for detailed refinery design (Khor et al. 2008; Khor and Elkamel 2010; Khor and Elkamel 2016; Albahri et al. 2018, 2019b) and for operation management (Pinto et al. 2000; Castillo Castillo et al. 2017; Ibrahim et al. 2018; Khor 2019a). A recent review on refinery optimization advances which encompass developments in both academic and industrial settings is available in Khor and Varvarezos (2017).

Industrial Case Study 1 attempts to contribute toward assessing heavy oil processing potential in the petroleum refining industry by adopting a model-based economic evaluation approach. Using product demands and crude oil feed properties as base data, a refinery model can be developed to evaluate potentially profitable technologies including those for residue oil upgrading. For this purpose, we formulate an optimization model suitable for a preliminary high-level investment decision-making with an appropriate economic objective function and a set of constraints that consider several conventional commercial technologies. A case study using available current data on market conditions is illustrated to carry out the intended assessment. A secondary goal of the study is to demonstrate the use of a standard business productivity tool (such as an Excel spreadsheet) to conduct such an assessment.

7.1.2 Problem Statement

We consider the investment decision-making problem for heavy oil processing in refineries given in the following data:

- fixed market demand for desired refinery products and their prices,
- available process technologies and their cost structures and capacities,
- cost of crude oil (single type or mixtures) and their nominal product yields.

We wish to determine the optimal process technologies or units and their indicative processing capacities (flow rates) by minimizing the total operating cost, which mainly consists of utility requirements on energy demand for processing operations.

7.1.3 Model Formulation

A refinery model suitable for preliminary investment decision-making is posed as an LP model. The model admits process parameters for heavy oil processing including raw material availabilities, nominal product yields of several representative commercial technologies, market demands and prices for main product streams, and global processing or product capacities besides various cost-related economic parameters. An optimum solution is determined as a point in the solution space that minimizes an economic-based objective function that stipulates the total operating cost for all heavy oil processing technologies considered that is feasible in satisfying all the associated constraints encompassing the said economic parameters.

This model uses the following notations:

Parameters:

$\mathrm{yd}_{i,u}$ Yield of component i from unit u,

oc_u Operating cost of unit u,

nc_i Capacity expansion cost of component i,

cp_u Capacity of process unit u,

dm_i Product demand of component i;

Variables:

$F_{i,j}$ Inlet flow rate of component i to unit u,

Z_i New capacity flow rate of component i.

A compact representation of the optimization model formulation is presented and explained in the following:

$$\text{minimize} \quad \sum_{i,u} \mathrm{oc}_u F_{i,u} + \sum_i \mathrm{nc}_i Z_i \tag{7.1}$$

$$\text{subject to} \quad \sum_u \mathrm{yd}_{i,u} F_{i,u} = 0, \quad \forall i \in I \tag{7.2}$$

$$\sum_i \mathrm{yd}_{i,u} F_{i,u} = 0, \quad \forall u \in U \tag{7.3}$$

$$\sum_i F_{i,u} \leq \mathrm{cp}_u, \quad \forall u \in U \tag{7.4}$$

$$Z_i + \sum_u F_{i,u} \geq \mathrm{dm}_i, \quad \forall i \in I \tag{7.5}$$

$$Z_i \leq \mathrm{av}_i, \quad \forall i \in I \tag{7.6}$$

$$F_{i,u}, Z_i \geq 0, \quad \forall i \in I \tag{7.7}$$

where the minimizing objective function shown in Eq. (7.1) caters for operating cost oc_u, which consists of raw material cost on crude oils and utility cost of process units based on their inlet flow rates as well as capacity expansion cost nc_i to meet market demands dm_i. Equation (7.2) describes component balances for each material i using fixed yield coefficients $yd_{i,u}$ (on mass basis) which render linear relation between the feed inputs and product outputs of unit u that are implicitly dependent on the unit's operating conditions. On the other hand, Equation (7.3) represents the total material balances for each unit u. Equation (7.4) ensures that the total inlet flows into unit u does not exceed its maximum capacity cp_u in determining the required processing level. Equation (7.5) stipulates that total processing rates for material i meet or exceed its demand dm_i including a provision for new capacity Z_i (or alternatively available product imports) to cover market requirements. Equation (7.6) specifies that the flow rate of imported component i does not violate its availability av_i (i.e. maximum amount) which also serves as an upper-bound constraint. Equation (7.7) enforces nonnegative values for all the decision variables.

7.1.4 Numerical Example

We consider a numerical example of assessing the worldwide potential for heavy oil processing in the downstream petroleum processing sector by applying the foregoing model. Economic model parameters are estimated based on commercial data available in the literature as cited in Tables 7.1–7.4. The raw material is assumed to be a vacuum residue stream available from a vacuum distillation unit or alternatively, a vacuum rerun unit with comparable processing capacity.

The products of each process technology are categorized according to their cut temperatures. Product yields of process technologies are typically given in volume percentages in the literature. To make use of the mass conservation principle, we convert them to weight percentages by assuming fixed densities of the product and feed components. The weight-based yields are then normalized (as listed in Table 7.2) according to the process technologies for use as input–output constants in the process unit material balances described by Eqs. (7.2) and (7.3).

Table 7.1 gives the product's economic parameters in terms of selling prices and market demands. The operating cost data for the heavy oil process technologies is

Table 7.1 Model economic parameters for products.

Product	Price (US$/kg)	Demand (kg/h)	References
Dry gas	0.0078	6,000	Gary et al. (2007) and Messick (2012)
Total LPG	0.0020	30,000	Gary et al. (2007) and Temizer (2017)
Gasoline	0.0097	150,000	Petrochemical Update (2017) and Maples (2000a)
Diesel	0.0039	121,000	Gary et al. (2007) and Fitzgibbon et al. (2018)
Gas oil	0.0031	40,000	Gary et al. (2007) and Fitzgibbon et al. (2018)
Coke	0.0027	60,000	Savant (2018)

Source: Khor (2019c). Reproduced with permission of IntechOpen.

Table 7.2 Product yields (in normalized weight percentages) for process units used in this work (Kamiya 1991; Gray 1994; Speight and Özüm 2002).

Product \ Technology	DCK	FCK	FCC	VB[a)]	VB[b)]	LCF	CP	FTC	HOT	HOC	MDS	ABC	HF	R2R
Dry gas	0.8	1.1	1.7	0.0	0.0	0.0	0.0	0.0	0.0	0.0	0.0	0.0	0.0	0.0
Total LPG	12.5	11.0	15.0	2.0	2.3	2.6	0.0	0.0	0.0	12.4	0.0	0.0	0.0	0.0
Naphtha	16.8	19.2	42.1	5.5	6.2	12.0	6.6	16.5	13.0	53.7	0.0	7.5	0.0	63.0
Middle distillate	14.9	22.9	20.6	10.5	11.8	41.2	30.7	38.2	9.7	14.6	61.2	17.9	4.2	18.1
Gas oil	33.6	13.1	10.1	82.0	79.7	44.2	62.7	24.2	28.7	8.8	38.8	74.5	25.5	4.8
Coke	21.5	32.7	10.4	0.0	0.0	0.0	0.0	21.1	48.6	10.6	0.0	0.0	70.3	14.1

a) Low API feed.
b) High API feed.
Source: Khor (2019c). Reproduced with permission of IntechOpen.

Table 7.3 Utility requirements (base data) and cost for heavy oil process technologies (Maples 2000a).

Technology	Electricity (kWh/b)	HPS (lb/b)	LPS (lb/bl)	Fuel (kBtu/b)	CW (gpm/b/h)	Cat (lb/b)	H_2 (ft^3/b)	Total (US$/(yr (kg/h)))
MDS	0.133	60	—	80	—	—	—	266.4
VB	0.033	−50	—	80	—	—	—	265.8
DCK	0.239	—	40	120	0.6	—	—	403.5
FCK	0.865	200	100	—	30	—	—	273.9
FCC	0.067	—	20	80	400	0.3	—	312.2
HOC	0.017	—	−80	80	—	0.25	—	267.0
HDT	0.093	—	—	24	400	—	900	128.3

Notes: HPS = HP Steam, LPS = LP Steam, CW = Cooling Water, Cat = Catalyst, H_2 = Hydrogen; unit b = barrel (0.136 barrel = 1000 kg), gpm = gallon/minute.
Source: Khor (2019c). Reproduced with permission of IntechOpen.

summarized in Tables 7.3 and 7.4. Annual operating time is taken to be 8150 hours per year corresponding to an onstream factor of about 93%.

The technologies considered in this case study (with their associated abbreviations in parentheses as used in Tables 7.3 and 7.4) are DCK; FCK; fluid catalytic cracking (FCC); VB; ebullated bed hydrocracking technology of LCF; Cherry-P (CP) and fluid thermal cracking (FTC) technologies; residual fluid catalytic cracking (RFCC) technologies of heavy oil treating (HOT), heavy oil cracking (HOC), and R2R (roughly stands for residue cracking with two-step regeneration); solvent deasphalting technology of MDS; and other residue hydrotreating (HDT) and hydroconversion technologies of asphaltenic bottoms cracking (ABC) and HF.

Furthermore, we consider several assumptions in representing operating requirements of these technologies. Solvent deasphalting operation depends on the solvent type. VB is a relatively inexpensive mild thermal cracking

Table 7.4 Operating cost parameters (Maples 2000a).

Parameter / Technology	Basis	Operating cost (US$/yr (kg/h))						
		SDA	VB	DC	FCK	FCC	HOC	HDT
Electricity tariff	0.060 US$/kWh	0.133	0.033	0.239	0.865	0.067	0.017	0.093
HP[a)] steam cost	0.0045 US$/lb	0.299	−0.249	0.200	1.330	0.100	−0.399	0.036
LP[b)] steam cost	0.003 US$/lb							
Fuel cost	3.000 US$/kBtu	266.0	266.0	399.0	0.0	266.0	266.0	79.8
Cooling water cost	0.10 US$/gal	0.000	0	3.990	3.325	44.336	0.000	44.34
Catalyst cost	5.00 US$/lb	0.000	0.000	0.000	0.000	1.663	1.386	0.000
Hydrogen cost	0.004 US$/ft^3	0.000	0.000	0.000	0.000	0.000	0.000	3.990
Total	—	266.4	265.8	403.5	273.9	312.2	267.0	312.2
Capacity (ton/h)	—	173.3	141.6	130.8	130.8	288.8	173.3	288.8

a) High pressure.
b) Low pressure.
Source: Khor (2019c). Reproduced with permission of IntechOpen.

process that is assumed to generate steam on a net basis, which can be sold (i.e. negative steam cost) while its cooling utility uses air instead of water. DCK requires a furnace to heat the feed stream for coke removal thereby it uses a large fuel quantity as compared to other technologies. FCK is a catalytic operation that uses steam for heating and air or water for cooling. HOC is similar to FCC with the capability to remove heat from the generator, which can be recovered to produce steam thus contributing as revenue (i.e. negative steam cost). HDT heavy oil consumes hydrogen in the reaction scheme to decrease carbon-to-hydrogen ratio, in which the model considers the worst-case operating requirements for cycle oil feed. Due to limited literature data, cost parameters for certain technologies are approximated to similar ones (e.g. LCF to FCK).

We use the model to conduct a general assessment of the probable technologies required to meet heavy oil processing capacity globally. The result obtained is graphically summarized in Figure 7.1. The objective value on total annualized cost of heavy oil processing is found to be about US$ 164.2 million with total utilities cost for the selected units determined to make up 80% while that of raw material cost only 17%. RFCC can account for nearly 31% of the available capacity, while the remaining can be met by a FCK technology. The potential RFCC technologies identified include HOC, HOT, and R2R (Kamiya 1991). It is also projected that there is demand for 53% of capacity expansion for heavy oil processing.

In general, RFCC can be designed compactly to produce high yields of valuable products with low maintenance costs as similar to the FCC technology that it is based upon. Fluid coker is reported to promote reactor heat transfer, which allows it to be operated at high temperature for high product yields with increased product separation into valuable products. It also uses burner operated with steam and air as

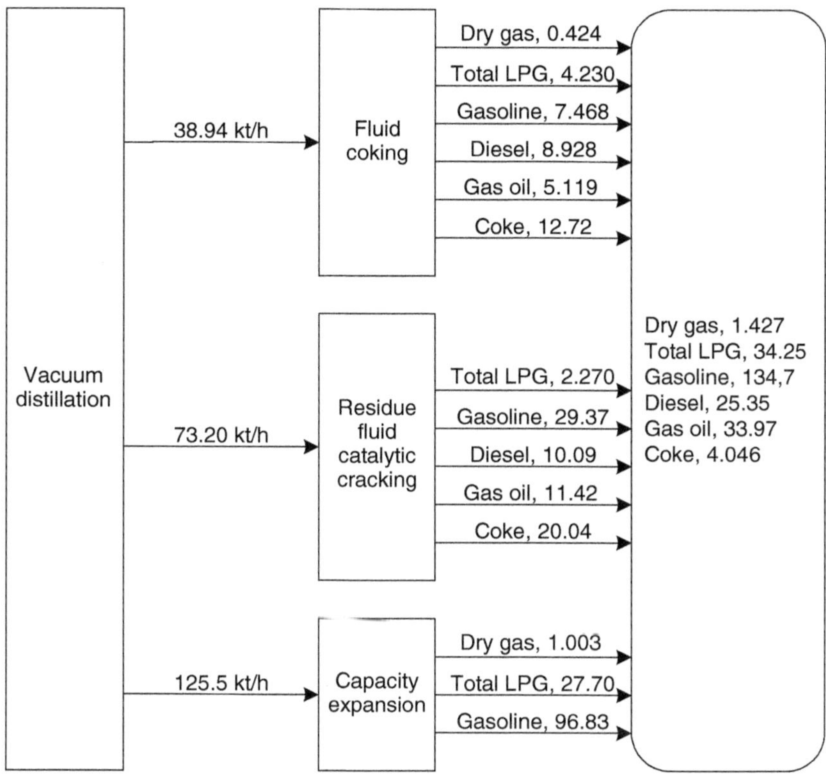

Figure 7.1 Model solution for the case study (all flow rates in kiloton per hour). Source: Khor (2019c). Reproduced with permission of IntechOpen.

utilities as opposed to an expensive fuel, which is reflected in its low operating cost (Speight and Özüm 2002; Speight 2011b).

We conduct sensitivity analysis to examine how the model parameter values influence the solution. As an example, Figure 7.2 shows the linear effect of varying the capacity expansion cost for diesel product output (in terms of a fixed multiplicative factor) on the total operating cost for heavy oil processing as is considered in our case study. Indeed, a trend of continuous high demand for distillate products (including diesel) necessitates correspondingly increased investment in processing cost.

7.1.5 Concluding Remarks

This chapter presents a model-based approach to conduct a preliminary assessment for investment decision-making in heavy oil processing for refineries. The economic evaluation can be carried out using an Excel spreadsheet or other similar business productivity tool. The results provide an order of magnitude indication of refining capacity potential for this increasingly important resource in the hydrocarbon industry.

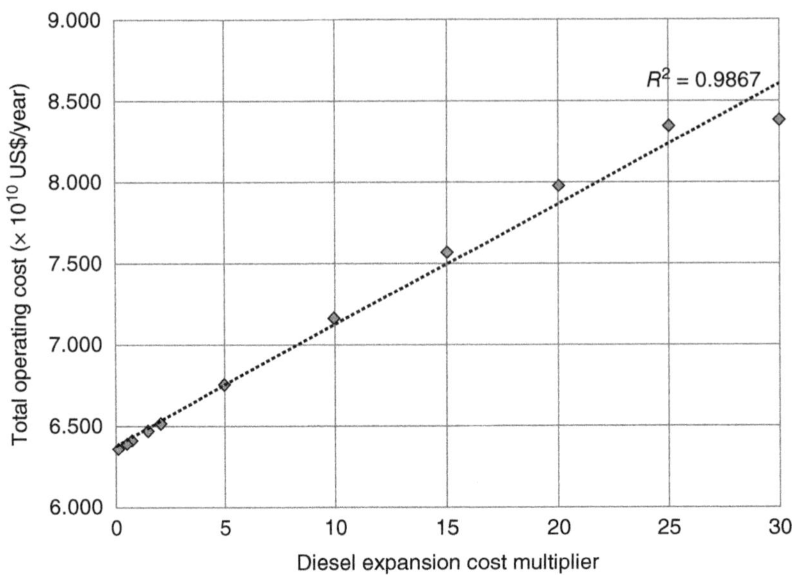

Figure 7.2 Sensitivity analysis on effect of diesel capacity expansion cost on total operating cost. Source: Khor (2019c). Reproduced with permission of IntechOpen.

7.2 Industrial Case Study 2: Refinery Configuration for Whole Complex Processing

This case study serves to exemplify that selecting an optimal configuration of processing sequence and intermediate stream routing becomes a challenging procedure when we consider the numerous heavy oil processing units with nearly a hundred commercial technologies and licenses available. This difficulty is due to the many possible alternatives arising from the combinatorial problem of sequencing any number of the process unit options. Thus, this case study contributes by developing a model-based optimization approach to determine an economically optimal refinery configuration without considering all possible refining schemes yet still meeting practical operating requirements. For this purpose, we present an aggregated network superstructure to preliminarily screen the numerous topology design alternatives to synthesize a large-scale grassroots refinery by accounting for existing main schemes for the desulfurization and cracking operations of crude oil mixtures. Based on this representation, we formulate a mixed-integer linear programming (MILP) model to perform structural and parameter optimization of the refinery configuration.

7.2.1 Model Formulation

Only one processing unit from each pool may be selected because they are mutually exclusive alternatives, i.e. only one unit is chosen amidst those that perform the same function. For instance, gas oil (GO) is processed in only one of the thermal, catalytic, or hydrocracking units (U_7 to U_{17}); we do not allow the model to combine two or more processes from the same pool. If the flow rate exceeds the process capacity, we use two trains of the same process to accommodate a large unit throughput as necessary. To stipulate this constraint, we incorporate additional logic propositions into the formulation as we further discuss next. These logical constraints reduce computational expense by providing information that increases the enumeration efficiency. They also ensure model integrity by incorporating qualitative design knowledge based on engineering experience and heuristics in refining process configuration. We do this by enforcing design specifications to select the units and streams linking the units besides structural specifications that associate the interconnectivity among the units by stipulating their relationships and describing the sequence of the streams linking the units.

Industrial Case Study 2 uses simplified correlations represented by linear equations mainly to represent product yields of process units, which is suitable for such technoeconomic study as applied in other work (Zhang and Zhu 2000, 2006; Zhang et al. 2001). Such overall and component material balances take the form of input–output flow rates with constant yields.

We use these relations to preserve model linearity for an MILP – the purpose is not to simulate a process unit or specific catalyst performance but to account for typical yields and properties in commercial operations. Such assumed linear models afford simplification and computational speed. Heavy oil processing correlations are obtained from Kamiya (1991) while those for lighter processes such as hydrogen production and purification, refinery gas processing, amine unit, sulfur production, catalytic reforming, and distillates (naphtha, kerosene, diesel, and GO) HDT are obtained from the cited sources (Gary and Handwerk 1994; Meyers 2016; Speight 2017).

We develop the model and implement the iterative computational procedure on a Visual Basic for Applications 7 platform running in Windows 7 environment on a Toshiba laptop with Intel Core i7 processor at 2.40 GHz of CPU clock speed and 8.00 GB of RAM. The constraints include material balances for the units and gathering pools for intermediate and final products as well as their possible interconnections. We also incorporate linear logical constraints using 0–1 binary variables to enforce certain design and structural specifications that tighten the formulation and enhance solution convergence.

We associate each process unit with a binary variable U_k, whose values alternate between 0 and 1 to indicate a unit's existence or non-existence in evaluating alternative refinery configurations. By changing such a structural variable value between 0 and 1, our modeling procedure exploits all conceivable structures that are embedded in the superstructure. The variables U_0–U_{95} are given in Figure 7.3 and Table 7.5 for their associated units in each processing pool. When their values alter between 0 and 1, we consider a new configuration and evaluate its profit function as given by selected unit capacities, flows of intermediate streams and final products, and incurred capital and running costs. If a new scheme is more profitable, we retain that configuration and its profit (i.e. the procedure stores values of the variables U_k and objective function). If not, we replace it with a next more profitable scheme until we

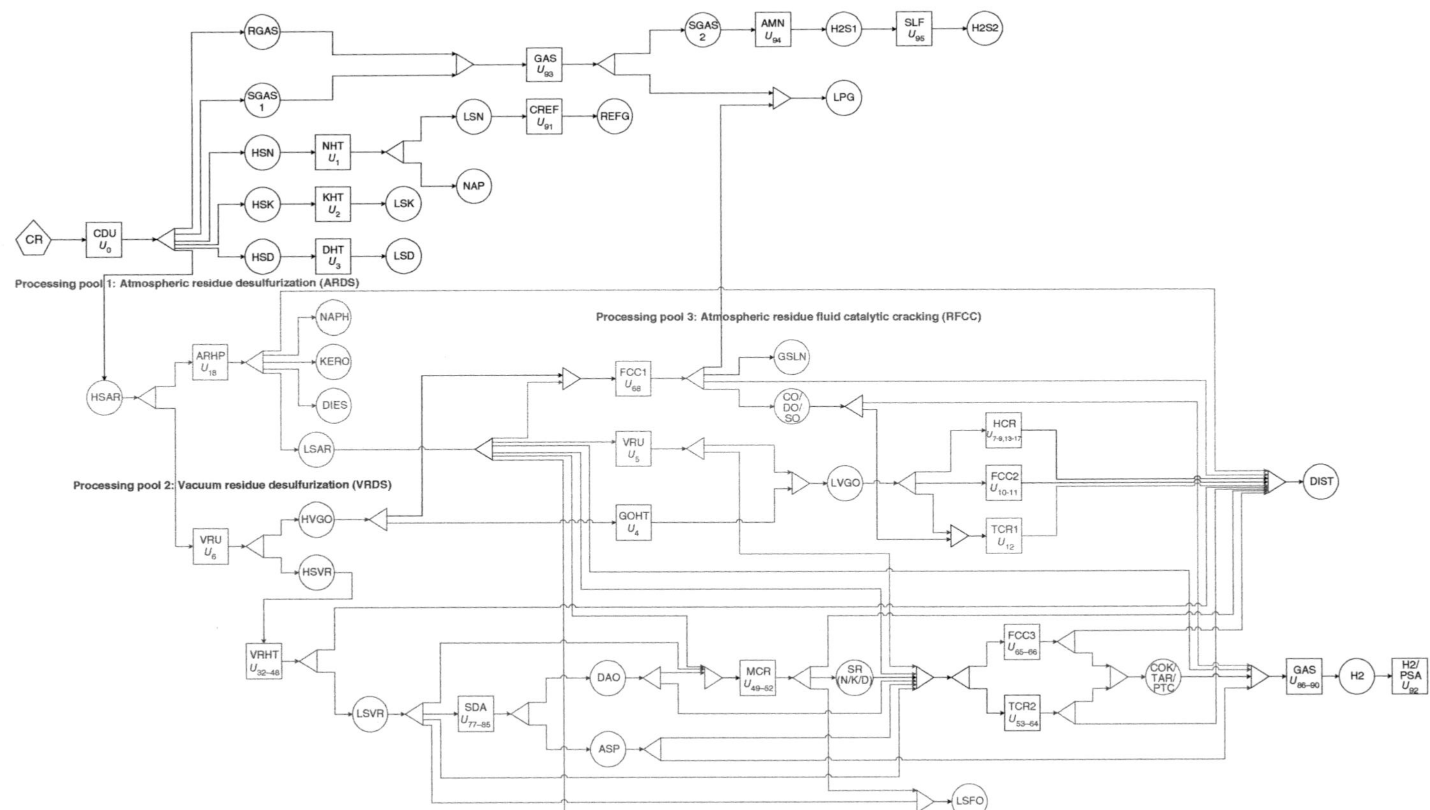

Figure 7.3 Aggregated superstructure representation of petroleum refinery configuration used in this study. Source: Albahri et al. (2018). Reproduced with permission of the American Chemical Society.

Table 7.5 Processing pools and/or operating modes of refinery process units are categorized in Industrial Case Study 2.

1. Gas oil cracking
- ACR
- Chevron Isocracking (gasoline, kerosene, and diesel modes)
- FCC
- IFP Hydrocracking
- S&W FCC
- UOP/UNOCAL Unicracking (gasoline, jet fuel, diesel, and gas oil modes)

2. Catalytic hydrotreating/hydroconversion of atmospheric residue
- CANMET
- Chevron ARDS/VRDS
- HDH
- HFC
- Hyvahl F
- LC-Fining
- MICROCAT–RC
- Residfining
- RHC
- RHC and GHC (combined processing)
- R-HYC (high and medium conversion modes)
- SHELL-RHC
- Unicracking/HDS

3. Mild cracking
- HYCAR
- TERVAHL-H
- TERVAHL-T
- Visbreaking

4. Solvent deasphalting
- DEMEX (low and high extraction levels)
- LEDA (30%, 53% and 65% extraction modes)
- MDS
- ROSE
- SOLVAHL (C4 solvent and C5 solvent modes)

5. Catalytic cracking of HSVR
- ART
- HOT

6. Catalytic cracking of LSAR
- APC FCC
- CMS-RFCC
- HOC
- R2R (gasoline and distillates modes)
- RCC (low, medium and high carbon residue)
- S&W RFCC
- Shell LR-FCC

7a. Catalytic hydrotreating/hydroconversion of vacuum residue
- CANMET hydrocracking
- Chevron RDS/VRDS hydrotreating (four product modes)
- HDH
- HFC
- H-Oil
- Hyvahl F (once through process)
- LC-Fining (high conversion – 90%)
- MICROCAT–RC
- MRH
- RCD UNIBON (BOC)
- RCD UNIBON (BOC) and thermal conversion
- Residfining
- SHELL-RHC Process
- SOC
- Unicracking/HDS (distillate mode)
- VCC
- VisABC

7b. Thermal cracking of vacuum residue
- ASCOT
- Cherry-P
- Delayed Coking (recycle ratios of 0.1, 0.3, or 1.08)
- ET-II
- EUREKA
- Flexicocking
- Fluid Coking
- FTC
- HSC
- KKI

8. Gasification of asphalt, pitch, tar, and coke
- Hybrid gasification
- Oxygen-blown KRW gasification
- Shell (partial oxidation)
- Texaco gasification
- Toyo THR-R

9. Common units
- Crude oil distillation
- Amine gas treating
- Catalytic reforming
- Claus sulfur plant (98% sulfur recovery)
- Naphtha hydrotreating
- Kerosene hydrotreating
- Diesel hydrotreating
- Gas oil hydrotreating
- Hydrogen production/Pressure swing adsorption
- Refinery gas processing
- Vacuum rerun (LSAR and HSAR modes)

Source: Albahri et al. (2018). Reproduced with permission of the American Chemical Society.

have evaluated all possible configurations through a branch-and-bound (Grossmann 2005) enumeration scheme. The number of alternatives is large, hence using logic propositions can reduce them to a reasonable yet meaningful quantity in devising an efficient computational procedure.

7.2.1.1 Superstructure Representation

We develop a superstructure in an aggregated form that aims to embed all possible alternative refinery configurations subsequently modeled using a MILP. Figure 7.3 shows a condensed version of the superstructure with the process units and streams denoted in the accompanying legend in Table 7.6. In this representation, the crude oil is first physically separated by a crude oil distillation unit (CDU) (U_0) into gases, LPG, naphtha, kerosene, diesel, and atmospheric residue (AR). Like the feed, CDU products have high sulfur and need treatment. The sour gas is typically treated for sulfur removal using a gas treating unit (U_{93}), amine unit (U_{94}), and sulfur recovery unit (U_{95}). Naphtha, kerosene, and diesel are desulfurized in separate catalytic hydrotreating units (HTU) (U_1–U_3) to lower the sulfur and other objectionable materials in final products to meet environmental regulations for sale. Naphtha is typically separated into light and heavy naphtha. Heavy naphtha can be converted into high-octane reformate gasoline blending stock in a catalytic reforming unit (U_{91}), while light naphtha can be isomerized to produce high-octane isomerate for gasoline blending. The light oil and gas processing sections of the refinery are standard with minor innovation challenge or improvement margin with not many alternatives to consider. On the other hand, the heavy oil processing section in a refinery (i.e. the bottom part of Figure 7.3) presents innovation opportunities that form the focus of our study.

The heavy oil processing technologies can be categorized into a few processing pools and/or operating modes as listed in Table 7.5. We develop each of the pools to comprise different processing technologies and their associated operating modes that perform similar functions (Kamiya 1991). Hence the selection of a pool is mutually exclusive from another and we devise our algorithmic procedure (using logic propositions; more explanation later) to select only one technology or mode from each pool.

7.2.1.1.1 Main Processing Alternatives

For the heavy oil portion of the refinery superstructure, we explore three main processing schemes to decide on an optimal route, namely, (1) atmospheric residue desulfurization (ARDS) alternative, (2) vacuum residue desulfurization (VRDS) alternative, and (3) atmospheric RFCC alternative. For the ARDS alternative, we first desulfurize high-sulfur atmospheric residue (HSAR) in one of the HDT and hydroconversion (hydroprocessing) units (U_{18}–U_{31}) to produce low-sulfur form, which is then physically separated in vacuum rerun unit to produce low-sulfur vacuum gas oil (VGO) and low-sulfur vacuum residue that can both be either sold or further processed to produce more valuable lighter products.

For the VRDS alternative, we physically separate HSAR in a vacuum rerun unit (U_6) into high-sulfur VGO and high-sulfur vacuum residue. We then desulfurize the

Table 7.6 Legend for symbols used in Figure 7.3.

State		Task	
CR	Crude oil	CDU	Crude distillation unit
HSN	Heavy straight-run naphtha	NHT	Naphtha hydrotreater
HSK	Heavy straight-run kerosene	KHT	Kerosene hydrotreater
HSD	Heavy straight-run diesel	DHT	Diesel hydrotreater
LSN	Light straight-run naphtha	CREF	Catalytic reformer
LSK	Light straight-run kerosene	GAS	Refinery gas unit
LSD	Light straight-run diesel	AMN	Amine unit
NAP	Light naphtha	SLF	Sulfur unit
NAPH	Low-sulfur naphtha	ARHP	Atmospheric residue hydroprocessor
KERO	Low-sulfur kerosene	FCC1	Catalytic cracker of low-sulfur atmospheric residue
DIES	Low-sulfur diesel	VRU	Vacuum rerun
LSAR	Low-sulfur atmospheric residue	HCR	Catalytic hydrocracker
REFG	Reformate gasoline	FCC2	Catalytic cracker of low-sulfur vacuum gas oil (LSVGO)
RGAS	Refinery gas	TCR1	Thermal cracker of LSVGO (includes advanced cracking reactor [ACR])
SGAS, SGAS2	Sour gas	GOHT	Hydrotreating of gas oil
H2S1, H2S2	Hydrogen sulfide	VRHT	Hydrotreating of vacuum residue
LPG	Liquefied petroleum gas	SDA	Solvent deasphalter
LSAR	Low-sulfur atmospheric residue	MCR	Mild cracker/(Hydro)Visbreaker
DIST	Distillate products	FCC3	Catalytic cracker of low-sulfur vacuum residue
CO/DO/SO	Cycle oil/Decant oil/Slurry oil	TCR2	Thermal cracker of high-sulfur vacuum residue (includes delayed coker)
LVGO	Low-sulfur vacuum gas oil (VGO)	GAS	Gasification
HVGO	High-sulfur vacuum gas oil	H2/PSA	Hydrogen production and pressure swing adsorption
HSVR	High-sulfur vacuum residue		
LSVR	Low-sulfur vacuum residue		
DAO	Deasphalted oil		
ASP	Asphalt		
SR(N/K/D)	Straight-run naphtha, kerosene, and diesel		
COK/TAR/PTC	Coke, tar, and pitch		
H2	Natural gas and refinery hydrogen-rich off gas		
LSFO	Low-sulfur fuel oil		

Source: Albahri et al. (2018). Reproduced with permission of American Chemical Society.

latter to get low-sulfur form in one of the vacuum residues HDT or hydroconversion units (U_{32}–U_{48}) with the dual intent to also produce more valuable products such as transportation fuels. High-sulfur VGO is converted in GO hydrotreater (U_4) to low-sulfur form.

For the RFCC alternative, we process low-sulfur AR in one of the RFCC units (U_{67}–U_{76}) to produce high-quality distillate fuels such as gasoline, jet fuels, and diesel. If cycle oil, decant oil, or slurry oil is produced, we can optionally sell or process it in a solvent deasphalting unit (U_{77}–U_{85}), mild cracker (U_{49}–U_{52}), thermal cracker (U_{53}–U_{64}), or gasifier (U_{86}–U_{90}) to produce either hydrogen or gas (the latter through gasification). We can see that each of these heavy oil processing options comprises numerous probable schemes since there are many alternative process units to choose from within each pool.

In the ARDS and VRDS alternatives, we upgrade low-sulfur VGO in a catalytic or thermal cracker or hydrocracker (U_7–U_{17}) to produce more valuable lighter products for sale. We sell high-quality light distillate products from AR and vacuum residue HDT/hydroconversion units (U_{18}–U_{48}) comprising liquefied petroleum gas, naphtha, kerosene, and diesel. Although some refiners sell low-sulfur vacuum residue as fuel oil, others further process it to produce more valuable lighter products such as using solvent deasphalter to produce deasphalted oil and asphalt, both of which can be sold or further processed in the refinery. We can feed the resulting asphalt to one of the gasification processes (U_{86}–U_{90}) to produce gas or hydrogen or process it by either thermal or catalytic cracking (U_{56}–U_{66}) to produce lighter distillate transportation fuels and as either coke or cracked residue (i.e. pitch and tar). We can also feed the cracked residue to one of the gasifiers (U_{86}–U_{90}). We can pretreat deasphalted oil in one of the mild thermal or hydro-VB units (U_{49}–U_{52}), send it directly to one of the gasifiers, or send it to one of the catalytic crackers or thermal crackers to produce light products for sale. Heavy bottoms produced from these residue (catalytic or thermal) crackers such as coke, pitch, or tar can be sold or further processed in one of the gasification units. Low-sulfur vacuum residue may also be processed directly in one of the catalytic or thermal crackers with or without pretreatment in a mild cracker to recover some light distillates.

In the proposed superstructure, solvent deasphalter (U_{77}–U_{85}) may precede, succeed, or completely replace vacuum rerun unit in the processing sequence. In one possible scheme, we first process HSAR in vacuum rerun unit to produce high-sulfur VGO and high-sulfur vacuum residue. This alternative requires not only desulfurizing high-sulfur VGO in GO hydrotreater and high-sulfur vacuum residue in HDT or hydroconversion but also to hydrotreat all products from solvent deasphalter, gasification, and catalytic and thermal cracking which have high sulfur if we do not treat vacuum residue first. Although we can use solvent deasphalter to prepare the hydrodesulfurization feed, deasphalting can also follow desulfurization to avoid a high sulfur in asphalt going to a delayed coker or gasification process or for sale. Likewise, we can use mild cracking processes such as VB, hydro-VB, or thermal cracking to pretreat a vacuum residue feed or product, but we can precede to desulfurize to obviate a high-sulfur product from these processes (Kamiya 1991; Meyers 2016).

We make special considerations for combination processes. A fluid thermal cracker (U_{62}) combines mild thermal cracking with coke gasification and therefore can neither be preceded by a mild thermal cracker (U_{49}–U_{52}) nor followed by a gasifier (U_{86}–U_{90}). An asphalt coking technology unit (U_{56}) combines deasphalting with thermal cracking and is neither preceded by a deasphalter nor followed by a thermal cracker. Similarly, a low-energy deasphalter (U_{77}–U_{79}) cannot precede asphalt coker nor can the latter coexist with a thermal cracker. They are mutually exclusive since asphalt coker combines deasphalting and coking. However, any of the following mild cracking technologies: HYCAR (U_{49}), VB (U_{50}), Tervahl-T (U_{51}), and Tervahl-H (U_{52}) can precede asphalt coking or thermal cracking.

We can thus see that the refinery superstructure constitutes a complicated interwoven web of process units and intermediate streams that is overwhelming. Any change in a process or a stream can result in structural and production pattern modifications, which require re-computing the optimal profit.

7.2.1.1.2 Modes of Operation

To model the different operating modes of a processing unit while maintaining system linearity, we model each mode as a separate unit. For example, the delayed coker is modeled as three distinct units (U_{53}–U_{55}), each with a different recycle ratio and material balance. We do the same for the gasoline, jet fuel, and diesel modes of operation for the isocracking process (U_7–U_9), and for the gasoline, jet fuel, diesel, and gasoil modes of operation for the UOP two-stage unicracking technology (U_{14}–U_{17}) (UOP 2010). We model the Axens IFP R2R low-sulfur atmospheric RFCC process as two units (U_{71} and U_{72}) for gasoline and distillate modes (Axens 2014). The vacuum rerun unit is modeled as two processes (U_5 and U_6) to account for high- or low-sulfur feed processing. The process units that the optimizer ultimately selects represent the preferred operating modes for each process within the entire refining scheme.

7.2.1.1.3 Decomposition-Based Superstructure

To enhance convergence, we devise a strategy that decomposes the superstructure into three processing pools as shown in a state-task network (STN)-based superstructure in Figure 7.3. Each pool represents a residual conversion refinery configuration corresponding to the three main alternatives for processing crude oil residue from the atmospheric distillation unit that we call the ARDS, VRDS, and RFCC schemes. Constructing a superstructure in this manner also helps to maintain relevance with practical refinery features – an optimal configuration is likely to resemble any one of the three schemes. In the following subsections, we discuss each of the sequences in terms of its topology, which is developed based on real-world existing refineries (Maples 2000a; Hydrocarbon Processing 2011; Meyers 2016).

ARDS Processing Scheme Figure 7.4 shows a residual conversion refinery topology based on an ARDS (AR desulfurization) scheme. It uses a crude flash separator followed by ARDS/VRDS hydrotreater (U_{18}) (e.g. the Chevron three product modes technology). We next perform vacuum flashing (U_5) of the low-sulfur AR to produce a delayed coker feedstock. Straight-run naphtha, kerosene, and diesel

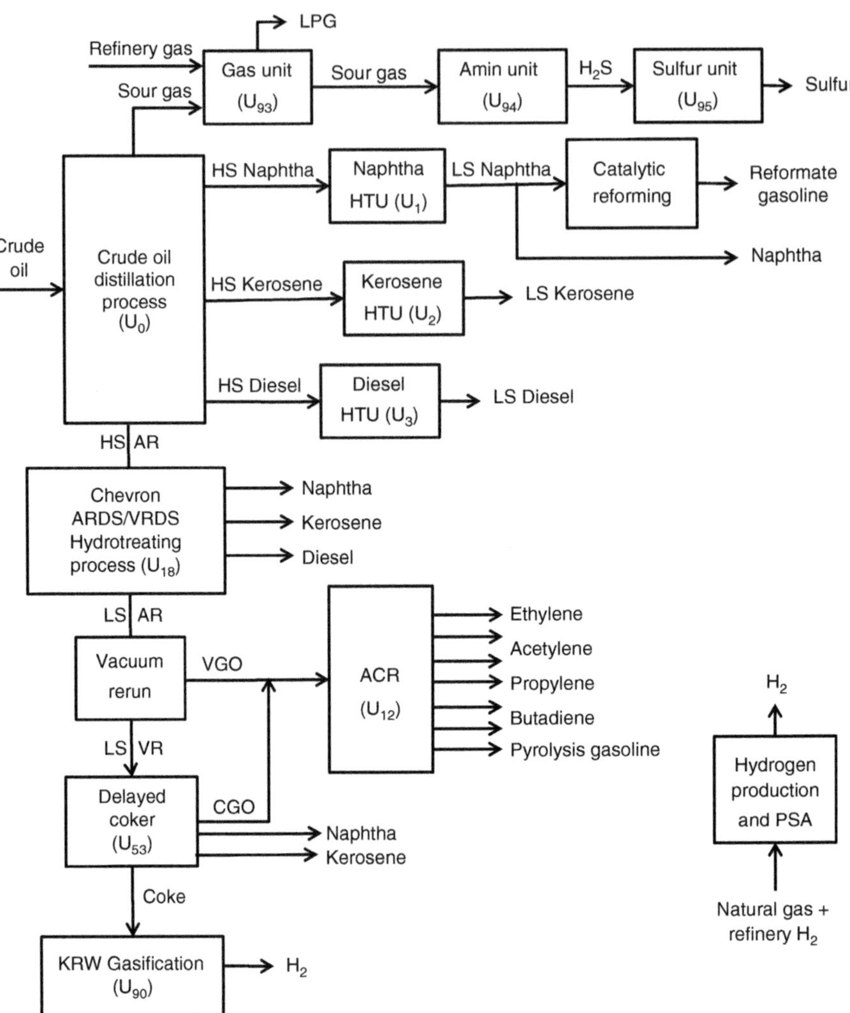

Figure 7.4 Simplified schematic representation of an ARDS refinery configuration, net refinery profit = US$ 24.53/bbl (58.40 cent/gal) of crude oil refined. Source: Albahri et al. (2018). Reproduced with permission of the American Chemical Society.

are hydrotreated. Then we split the hydrotreated naphtha into light and heavy fractions for which the latter is catalytically reformed to improve octane rating. Kerosene is hydrotreated to reduce its aromatic and naphthalene content to meet jet fuel specifications. We convert the low-sulfur vacuum and coker gas oils (CGOs) in an advanced cracking reactor (ARC; U_{12}) to yield ethylene, acetylene, propylene, butadiene, and pyrolysis gasoline. This option is more profitable than using conventional GO FCC or hydrocracker. Delayed coker converts vacuum residue to cracked distillates and green coke. Using ARDS upstream of the delayed coker improves the ultimate coke quality by reducing sulfur and metals in the coker feed. To compare, low-sulfur GOs produced via this scheme give better ARC yields than untreated CGOs from a straight-run naphtha residue. Because ARDS reduces the carbon residual content of coker feed, coke yield from delayed coker is less than that of untreated reduced crude oil feed (Caballero and Grossmann 1999). A vacuum rerun unit downstream of ARDS further reduces coke yield. Moreover, operating delayed coker at 10% recycle ratio is more profitable. Hence it is advantageous to convert coke to hydrogen in KRW gasification process (U_{90}) to supplement refinery needs.

VRDS Processing Scheme Figure 7.5 shows another type of residual conversion refinery topology that is based on a VRDS scheme. It initially performs vacuum

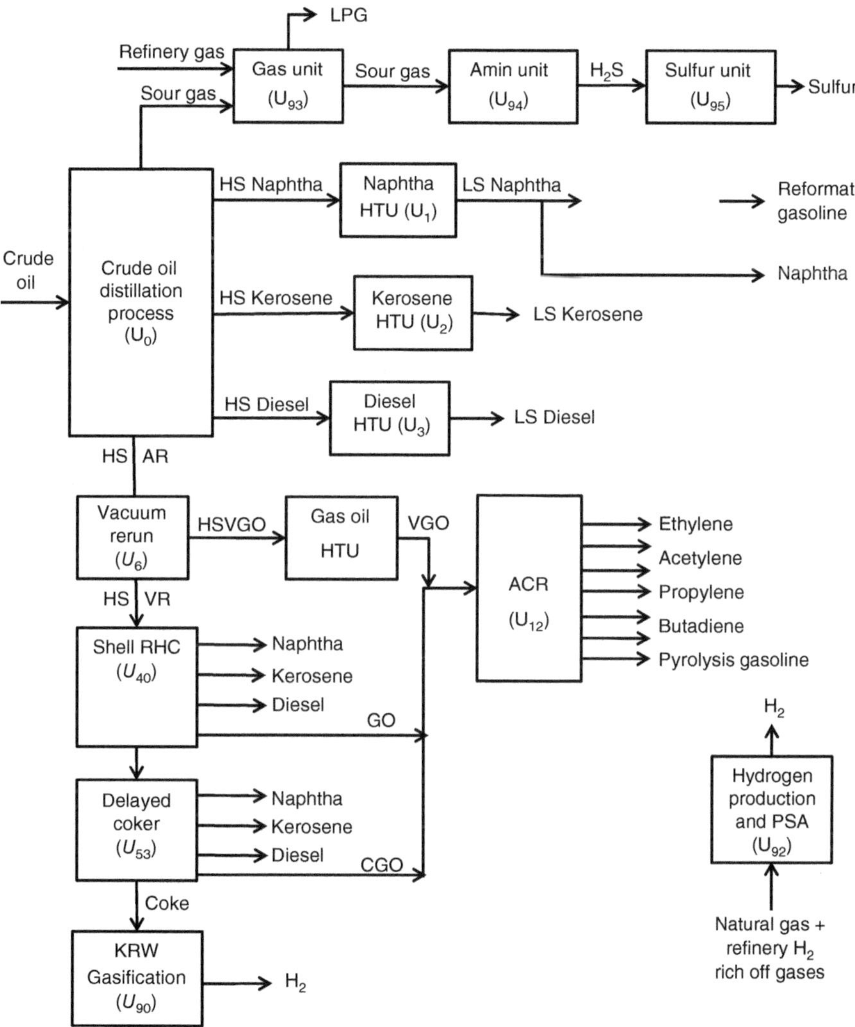

Figure 7.5 Simplified schematic representation of a VRDS refinery configuration, net refinery profit = US$ 22.5/bbl (54.20 cent/gal) of crude oil refined. Source: Albahri et al. (2018). Reproduced with permission of the American Chemical Society.

flashing of HSAR. This configuration uses a combined crude and vacuum flasher coupled with GO hydrotreater and advanced cracker to hydrotreat straight-run naphtha, kerosene, and diesel. Like ARDS scheme, we then split the hydrotreated naphtha into light and heavy fractions to catalytically reform the latter while HDT straight-run kerosene to jet fuel. High-sulfur vacuum residue is desulfurized by Shell RHC catalytic hydrotreater/hydroconverter (U_{40}) followed by DCK at 10% recycle ratio (U_{53}). Low-sulfur GO from vacuum rerun unit, delayed coker, and Shell RHC is fed to ARC, which as is similar to the ARDS scheme, is found to be feasible and preferred over using conventional hydrocracker or FCC. ARC converts low-sulfur GO to high yields of ethylene, acetylene, propylene, butadiene, and hydrolysis gasoline. Delayed coker converts vacuum residue to cracked distillates and green coke. Since Shell RHC reduces the coker feed's carbon residual content, coke yield from delayed coker is less than that of an untreated reduced crude feed.

RFCC Processing Scheme Figure 7.6 shows a residual conversion refinery based on RFCC scheme. It involves residual cracking of low-sulfur AR using crude flash followed by AR desulfurization and then RFCC. We desulfurize reduced crude by the Chevron process followed by FCC (U_{68}) with advanced process control (APC) to catalytically crack low-sulfur AR. Straight-run naphtha, kerosene, and diesel are hydrotreated. As is similar to the previous configurations, we then perform the following operating sequence: (1) split the hydrotreated naphtha into light and heavy fractions to catalytically reform the latter; (2) hydrotreat kerosene to jet fuel; (3) convert low-sulfur AR in U_{68} to liquefied petroleum gas, gasoline, light cycle oil, and decant oil; and (4) crack light cycle oil in ARC, in which this option is again preferred to hydrocracker or FCC to get high yields of ethylene, acetylene, propylene, butadiene, and hydrolysis gasoline. Besides improving FCC product yields, desulfurizing residue results in lower sulfur emissions from FCC regenerator and allows using FCC decant oil as low-sulfur fuel oil. Note that a low-sulfur light cycle oil produced via this route gives better ARC yields than an untreated one from a straight-run residue.

7.2.1.2 Logic Propositions

We optimize the selection of process units and technologies by using a set of binary variables defined as follows:

$$U_k = \begin{cases} 1, & \text{process unit } k \text{ is selected} \\ 0, & \text{process unit } k \text{ is not selected} \end{cases} \quad k = 1, 2, \ldots, 96 \tag{7.8}$$

We enforce certain process units such as a crude distillation unit (CDU, U_1) must exist since a refinery cannot operate without them. The same is true for U_1–U_3 to desulfurize naphtha, kerosene, and diesel products, and for U_{91}–U_{95} to ensure sulfur removal, high octane number, and hydrogen supply:

$$U_k = 1, \quad k = 0, 1, 2, 3, 91, 92, 93, 94, 95 \tag{7.9}$$

A vacuum rerun unit (U_5 and U_6) cannot coexist because they represent two modes of operation for the same process unit as described by:

$$U_5 + U_6 \leq 1 \tag{7.10}$$

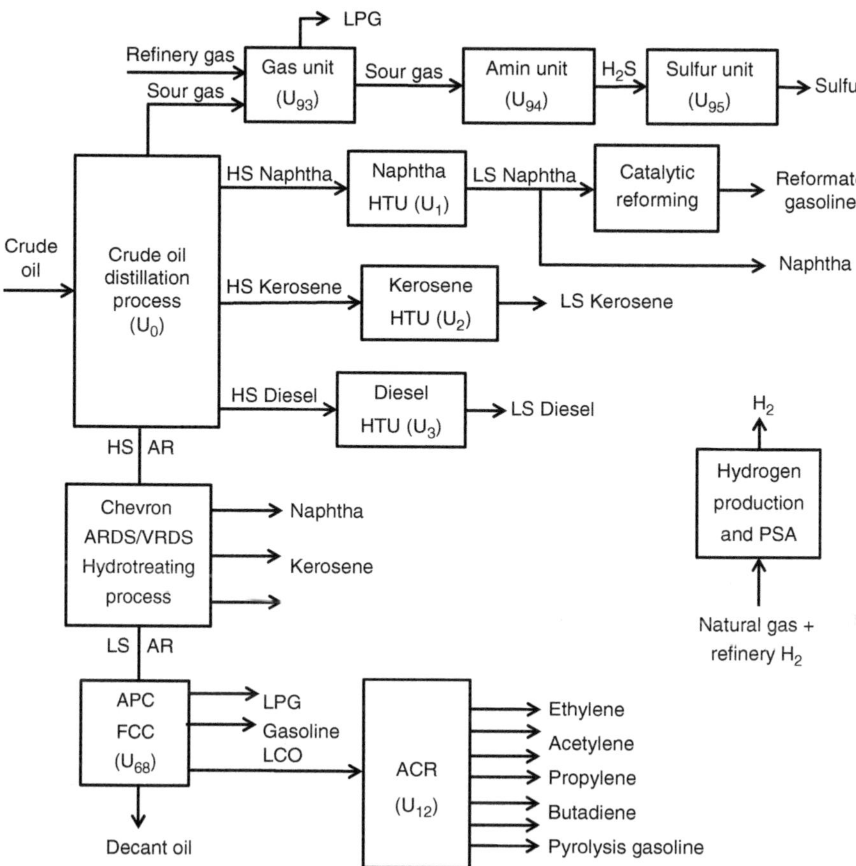

Figure 7.6 Simplified schematic representation of RFCC configuration, net refinery profit = US$ 25.55/bbl (60.8 cent/gal) of crude oil refined. Source: Albahri et al. (2018). Reproduced with permission of the American Chemical Society.

Likewise, AR (U_{18}–U_{31}) and vacuum residue (VR, U_{32}–U_{48}), HDT or hydroconversion units are mutually exclusive and cannot coexist because desulfurizing AR or VR negates a need to do so for the other. The hydrotreater (U_{18}–U_{31}) exists only if we produce HSAR; similarly, for the latter (U_{32}–U_{48}). We do not process HSAR in both high-sulfur vacuum rerun unit and ARDS. That means, U_6 cannot coexist with any of U_{18}–U_{31}; if any of $U_{18} - U_{31} = 1$, then $U_6 = 0$ and vice versa, i.e. $\forall U_k = 1 \Leftrightarrow U_6 = 0$ for $k = 18, 19, \ldots, 31$. The logical constraint in Eq. (7.11) describes this condition:

$$U_k + U_6 \leq 1, \quad k = 18, 19, \ldots, 31 \tag{7.11}$$

Similarly, low-sulfur vacuum rerun unit (U_5) cannot coexist with any of U_{32}–U_{48}, which means any of $U_{32} - U_{48} = 1$ implies $U_5 = 0$ and the converse is true, i.e. $\forall U_k = 1 \Leftrightarrow U_5 = 0$ for $k = 32, 33, \ldots, 48$:

$$U_k + U_5 \leq 1, \quad k = 32, 33, \ldots, 48 \tag{7.12}$$

But GO hydrotreater (U_4) exists only if we produce high-sulfur GO (i.e. $F_{HSGO} > 0$) from vacuum rerun unit (U_6):

$$U_4 \leq U_6 \tag{7.13}$$

The catalytic cracking and thermal cracking of residual oil handle the same feed type; hence they cannot coexist or are mutually exclusive. Thus, if we select a process technology from a catalytic or thermal cracking pool, we do not need another. Likewise, RFCC catalytic cracking of low-sulfur AR (U_{67}–U_{76}) cannot coexist with U_5 because they have the same feed and the converse applies, i.e. $\forall U_k = 1 \Leftrightarrow \forall U_l = 0$ for $k = 53, 54, \ldots, 64$, $l = 65, 66, \ldots, 76$:

$$U_k + U_l \leq 1, \quad k = 53, 54, \ldots, 64, \quad l = 65, 66, \ldots, 76 \tag{7.14}$$

In addition, when any of U_{67}–U_{76} exists, U_6 does not exist because these two options constitute two alternative routes.

A unit exists only if it has a feed; if no downstream unit exists, we sell the feed stream as a product. To illustrate, GO conversion units (U_7–U_{17}) exist only if we produce GO. If no GO conversion unit exists ($\forall(U_7 - U_{17}) = 0$), then low-sulfur GO is sold. Another example is if both solvent deasphalter products, i.e. deasphalted oil and asphalt go to the same destination unit, then we eliminate solvent deasphalter because it is suboptimal to separate its feed and combine it again (i.e. $U_{77} - U_{85} = 0$).

7.2.1.3 Objective Function

The economics-based objective function of our model includes feed and product prices, utility requirements, and running costs for the units. We account for investment or capital cost that includes the process unit erection cost and paid-up royalty fees as applicable (Khor and Elkamel 2010). We do not include erection cost of supporting units such as steam generation, wastewater treatment, offsites, tank farm, or shipping and loading facilities. Moreover, our investment cost structure does not cover working capital, start-up expense, inventories, or cost of land, site preparation, taxes, licenses, permits, and duties. We assume these costs are constant for all refinery configurations, hence they do not affect an optimal solution because we compare how the alternatives differ in incremental terms and not by total investment.

We consider operating or running cost that includes running-royalty fees, payroll, maintenance, insurance, taxes, purchased catalysts, chemicals, and utilities. We purchase all utilities required covering steam, electric power, fuel, cooling water circulation, boiler feed water, and process injection water. Running cost does not include charges for product delivery or blending and neither that of chemicals such as oxygenates. For simplicity, we assume labor cost to be the same for all refinery configurations.

We emphasize here that our interest is not to evaluate the total investment cost rigorously but rather to compare the relative merits of the alternatives. The full details of the cost data and correlations that we adopt are reported including their sources and justifications for use. It is noteworthy that the nonlinear cost relations in the model involve fixed-value parameters and not variables (hence the model linearity is preserved).

The complete model formulation is as follows with an objective function of maximizing the total refinery profit:

$$\max \sum_i \text{sp}_i P_i - \sum_j \text{cf}_j F_j - \sum_k (\text{CC}_k + \text{CR}_k) - \text{CM} - \text{CP}$$

$$\text{s.t. } A\mathbf{x} = 0$$

$$By \leq b \tag{7.15}$$

where sp_i is unit sale price of product i, P_i is flow rate of product i, F_j is flow rate of feed j, cf_j is unit purchase price of feed j (mainly crude oil and natural gas), CC_k is capital cost (amortized) of process unit k, CR_k is running cost of k, CM is cost of maintenance, tax, and insurance, and CP is total payroll cost. $A\mathbf{x} = 0$ represents constant-yield correlation-based material balances where A is a matrix of linear constant yields and $\mathbf{x}$ is a decision variables vector of material flow rates. The linear constraint $B\mathbf{y} \leq b$ represents a logic proposition.

7.2.2 Computational Results

We apply our proposed modeling approach to an actual medium size grassroots refinery called Mina Abdullah that Kuwait National Petroleum Company (KNPC) owns and operates. The facility processes 200,000 bpd of a crude oil mixture for export with a specific gravity of 30 API and sulfur content of 2.6 wt% to produce transportation fuels. The case study considers the following operational features:

- Refinery gas is treated in a gas unit, amine unit, and finally in a sulfur production unit;
- Hydrogen requirement is met by producing using purchased natural gas and supplemented from the reformer and hydrogen-rich off-gases in hydrogen recovery unit;
- Required fuel and utility are met by external purchase while produced gas and utilities such as steam are sold;
- Normal butane is blended with gasoline to maximize profit;
- Combined streams of propane-and-butane are sold as liquefied petroleum gas (LPG);
- Gas streams containing mixed methane-and-ethane or methane-to-butane are not separated in a gas plant; they are priced using free-on-board (FOB) for sale;
- No alkylation or isomerization unit exists;
- Erection cost is amortized over a 50-year operating life;
- Only low-sulfur products are sold whereas high-sulfur products are processed.
- Streams with boiling points of about 370 °C and above (370+ °C) produced from units other than CDU are categorized as long residue while those with boiling points of 500+ °C from units other than vacuum rerun unit are categorized as short residue.

The feed and product prices are given in Table 7.7 and utility prices in Table 7.8. CM is taken as 3% of the total annualized erection cost (Caballero and Grossmann 1999) and CP accounts for 1500 employees.

Table 7.7 Refinery feed and product prices for Q3 2014 (in US$) (Rabiei 2012; Meyers 2016; Fibre2Fashion Pvt Ltd 2017; FillLPG.co.uk 2017; IndexMundi 2017; QPSPP 2017; SunSirs-China Commodity Data Group 2017; U.S. Energy Information Administration 2017).

Product	US$/bbl	US$/ton
Crude oil (Brent June 2014)	108	776
Liquefied petroleum gas	110	1260
Hydrotreated naphtha	113	1013
Gasoline	121	1085
Hydrotreated kerosene	120	944
Hydrotreated diesel	122	905
Gas oil product	112	756
High-sulfur atmospheric residue	90	587
Low-sulfur atmospheric residue	96	628
High-sulfur vacuum residue	90	567
Low-sulfur vacuum residue	96	607
Pitch	131	620
Asphalt	94	515
Heavy fuel oil	117	744
Light fuel oil	118	904
Natural gas (US$/MBtu)	3.75	—
Hydrogen (US$/$10^6$ cm^3, US$/MMSCF)	544	—
Petroleum sulfur	—	166
Petroleum coke (calcined)	—	135
Methanol	—	220
Ethylene	—	750
Acetylene	—	860
Propylene	—	1300
Crude butadiene	—	3300
Ammonia	—	460

Source: Albahri et al. (2018). Reproduced with permission of the American Chemical Society.

7.2.2.1 Computational Results and Discussion

The model with logic propositions considers a total of 2,164,500 configurations, which would be the plot plans evaluated in a non-model-based study using heuristics. The details on the configurations for each scheme are shown in Table 7.9 together with its net product yields (in ton/day or bpd as appropriate) and associated cost and profit. Simplified block diagrams for the configurations are shown in Figures 7.8–7.10. Next, we analyze the results from evaluating the configuration alternatives.

Table 7.8 Utility prices (Meyers 2016).

Utility	Value	Unit
High pressure steam	0.0041	US$/lb
Medium pressure steam	0.00365	US$/lb
Low pressure steam	0.0031	US$/lb
Power	0.08	US$/kWh
Cooling water	0.00005	US$/gal
Fuel (net)	2.02	US$/MBtu
Boiler feedwater	0.0063	US$/gal
Process/injection water	0.0054	US$/gal
Condensate	0.0054	US$/gal
Hydrogen	850	US$/$10^6$ cm^3
Monoethanolamine	6.78	US$/gal
(MEA) solution	0.80	US$/lb
Oxygen	0.19	US$/ton
Maintenance allowance	2% of erection cost	US$/year
Taxes and insurance	1% of erection cost	US$/year

Source: Albahri et al. (2018). Reproduced with permission of the American Chemical Society.

An optimal configuration that we obtain in our computational example as shown in Figure 7.7 reveals that it is economically optimal to build RFCC and ACR within a single site as economics permit. In this scheme, first, we physically separate crude oil in CDU into gases, LPG, naphtha, kerosene, diesel, and AR. Then we desulfurize straight-run naphtha, kerosene, and diesel in separate hydrotreaters to produce low-sulfur products for sale. Past experience indicates to first desulfurize the HSAR in a hydrotreater/hydroconverter. Then we catalytically crack instead of vacuum flash the resulting low-sulfur AR in FCC to produce LPG, gasoline, light cycle oil, and decant oil, which is consistent with industrial practice reported in Maiti et al. (2001) Although we can process the residue in a thermal cracker or catalytic cracker to produce more valuable lighter products (after using one or more of vacuum flash, solvent deasphalter, or mild cracker), this route is not preferred. The model selects ACR (thermal cracking process of LSGO) in all three ARDS, VRDS, and RFCC cases instead of catalytic cracking or hydrocracking. Hydroprocessing (Chevron technology) is optimal to treat HSAR; using integer programming technique (by applying a linear integral constraint to restrict options) (Biegler et al. 1997) shows that UOP unicracking technology is the next preferred choice. Distillate products from hydroprocessing, namely, LPG, naphtha, kerosene, and diesel are sold.

Table 7.9 lists the total refinery investment, detailed operating expense requirements, and net profit. Estimated total capital investment ranged from US$ 1.755 to US$ 1.918 billion, which is consistent with reported values for a refinery of

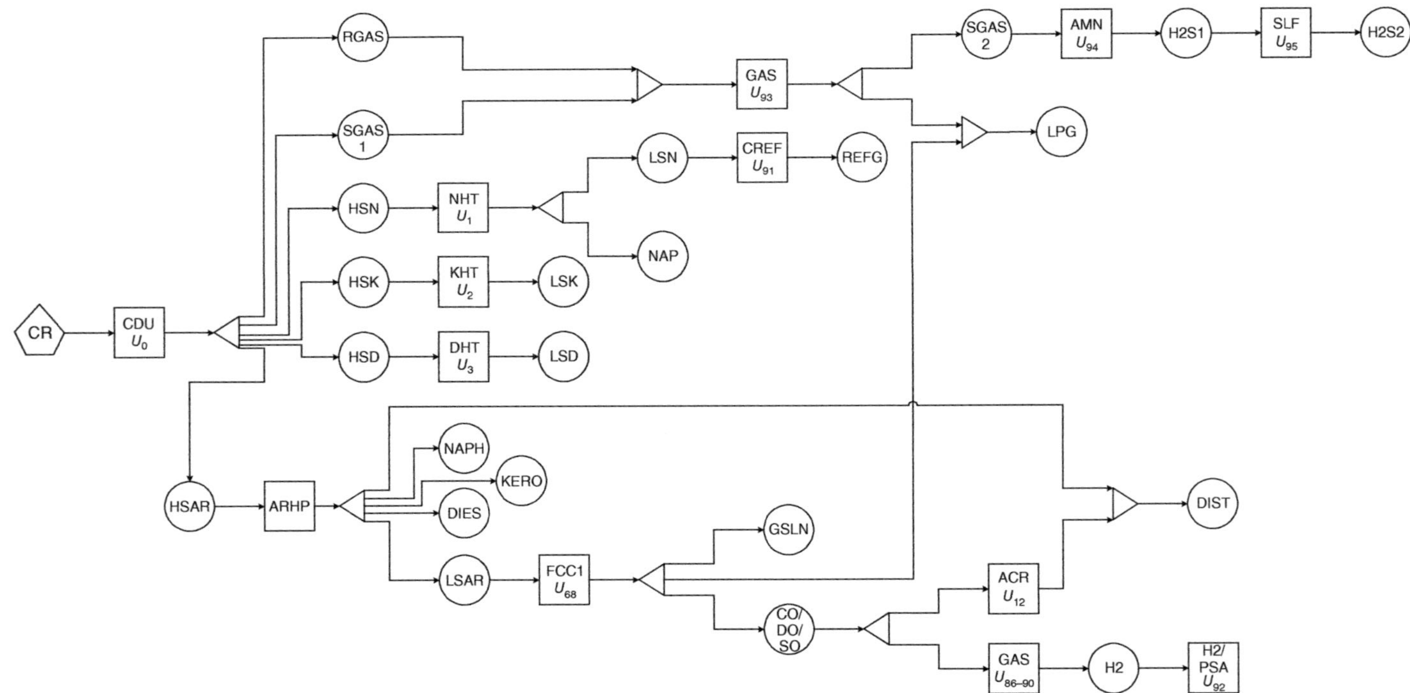

Figure 7.7 Optimal solution of a residual conversion refinery configuration in the computational example. Source: Albahri et al. (2018). Reproduced with permission of the American Chemical Society.

Table 7.9 Refinery net product yields and associated economic evaluations for residue conversion refinery configurations with 200,000 barrel per day (bbl/d) capacity.

	VRDS configuration		ARDS configuration		RFCC configuration	
Product	**bbl/d**	**ton/d**	**bbl/d**	**ton/d**	**bbl/d**	**ton/d**
Crude oil (Brent June 2014)	−200,000	−30,120	−200,000	−30,120	−200,000	−30,120
Natural gas (ft^3/d) (scfd)	−20,906,260	−419	−19,013,065	−395	−28,176,315	−585
Refinery gas, in fuel oil equivalent (FOE)	6,890	4,621	7,433	4,232	3,206	3,827
Liquefied petroleum gas	2,900	252	2,900	252	29,770	2,591
Naphtha	28,034	3,127	34,557	3,854	32,554	3,631
Gasoline	31,977	3,567	33,607	3,748	72,520	8,089
Kerosene	27,820	3,542	27,820	3,542	27,820	3,542
Diesel	45,365	6,123	48,008	6,480	48,008	6,480
Heavy fuel oil	9,584	1,501	9,399	1,472	2,284	358
Light fuel oil	9,499	1,245	9,315	1,221	2,264	297
Methanol	418	52	1,350	169	0	
Petroleum sulfur		720		720		720
Petroleum coke (calcined)		0		0		0
Ethylene		2,853		2,513		611
Acetylene		178		156		38
Propylene		1,300		1,145		278
Crude butadiene		1,141		1,005		244
Total capital cost (US$)	1,755,413,153		1,918,056,927		1,837,085,416	
Capital recovery (0%/d, 50 y, 365 d/y)	96,187		105,099		100,662	
Running cost (US$/d)	496,730		421,688		412,365	
Maintenance allowance, taxes, and insurance (US$/d)	144,281		157,649		150,993	
Total personnel[a] (US$/d)	550,000		550,000		550,000	
Natural gas cost (US$/d)	71,147		66,914		99,163	

(continued)

Table 7.9 (Continued)

Product		VRDS configuration		ARDS configuration		RFCC configuration	
		bbl/d	ton/d	bbl/d	ton/d	bbl/d	ton/d
Crude oil cost (US$/d)		21,600,000		21,600,000		21,600,000	
Total sales from products (US$/d)		27,508,767		27,808,064		28,023,699	
Net refinery profit,	US$/yr	1,660,903,883		1,790,950,808		1,865,338,148	
	US$/day	4,550,422		4,906,715		5,110,515	
	US$/bbl	22.75		24.53		25.55	
	cent/gal	54.20		58.40		60.80	
Internal rate of return (%)		9.9		9.7		10.6	
No. of refinery schemes[b)]		841,500		693,000		630,000	

a) Number of employees = 1,500; average daily pay per employee = US$ 300; total daily payroll = US$ 450,000; estimated daily bonus, benefits, training, and advertising = US$ 100,000.
b) Total number of refinery process schemes evaluated = 2,164,500.
Source: Albahri et al. (2018). Reproduced with permission of the American Chemical Society.

comparable size (Axens 2014). The optimal configuration is the RFCC scheme as shown in Figure 7.7 with a net profit of US$ 25.55/bbl of crude oil refined, which is equivalent to 60.83 cent/gallon (cent/gal). Using the aforementioned integer programming technique of appending an integral cut to remove routes associated with the RFCC alternative, we get the ARDS scheme as the next preferred configuration followed by VRDS. Nevertheless, the RFCC scheme is only marginally more profitable than the ARDS scheme and involves relatively fewer units. The internal rate of return based on a discounted cash flow calculation for each of the residual conversion refinery schemes is also computed – importantly, the trend of values agrees with our result.

Figure 7.7 shows the specific process technologies that the model selects from each pool for the RFCC configuration. It reveals that HDT straight-run products is favored. HDT straight-run distillates from CDU is economically optimal than selling them as straight-run high-sulfur products which fetch lower prices. To illustrate, HDT straight-run naphtha only costs about US$ 0.49/bbl, whereas its price compared to in hydrotreated form is about US$ 6.00–US$ 7.00/bbl lower. Our computational study gives solution runs that consistently involve HDT naphtha, kerosene, and diesel, hence we decide to incorporate such structure in the model through logic propositions.

RFCC of AR is favored to thermal cracking due to the catalytic feature of the former that can be tailored to improve selectivity and yield. But advanced thermal cracking of GO is favored to catalytic cracking or hydrocracking because its product pattern is more valuable and suitable for petrochemicals. Unless preceded by AR HDT, RFCC removes only 30–50% of sulfur and its products still need HDT; likewise, products from other hydrocracking technologies (e.g. HDH, HFC, and MRH) still have high sulfur which requires further HDT. Consequently, these routes are not considered optimal.

Table 7.9 reports the respective erection and total running costs (including maintenance and payroll) for the RFCC scheme as 1.20 and 13.30 cent/gal, respectively, totaling 14.50 cent/gal. On the other hand, total costs for optimal ARDS and VRDS schemes are 14.69 and 15.50 cent/gal, respectively. Their relatively marginal cost differences underline the significance of employing the logic propositions to arrive at an optimal refinery configuration. In addition, capital cost impacts less on the total cost as compared to running cost. To illustrate, if a refinery is located in a remote place such as Alaska, erection cost triples (Khor and Elkamel 2010) but reduces the RFCC scheme profit by only 3.60 cent/gal, translating to only a small impact since capital cost triples correspondingly for the other schemes too. Therefore, a refinery configuration is relatively more sensitive to product and feedstock prices than capital cost and to a lesser extent on running cost.

The optimal profit (corresponding to RFCC scheme) of 60.80 cent/gal is consistent with the higher range of an average US refinery profit of 30–60 cent/gal. We note the presence of uncertainty due to our product prices and erection and operating cost estimates besides the excluded cost components (Khor and Elkamel 2010). In part, we mitigate the uncertainty by relying on multiple data sources to compare trends over time periods and relations with crude oil prices in arriving at a representative average price to use in our computational study. For instance, Table 7.7 lists product prices in terms of the wholesale prices and not the typically higher final consumer prices as these prices can be volatile and fluctuate even when crude oil price is constant. This way, it ensures our study findings remain consistent despite significant changes in crude oil price.

The computational statistics of our modeling study is summarized in Table 7.10.

7.2.2.2 Model Validation

To validate the model and illustrate the applicability of our proposed approach, we compare the computational results obtained for the Mina Abdullah (MAB) refinery to other refinery configurations operated by KNPC, namely, that of Shuaiba (SHU)

Table 7.10 Model size and computational statistics.

Model type	MILP
Computing platform	Microsoft Excel 2010
Solver	Excel Solver (Frontline Systems) (Frontline Systems 2018)
No. of continuous variables	574
No. of 0–1 variables	110
No. of constraints	112
CPU time	<0.1 s (trivial)

Source: Albahri et al. (2018). Reproduced with permission of the American Chemical Society.

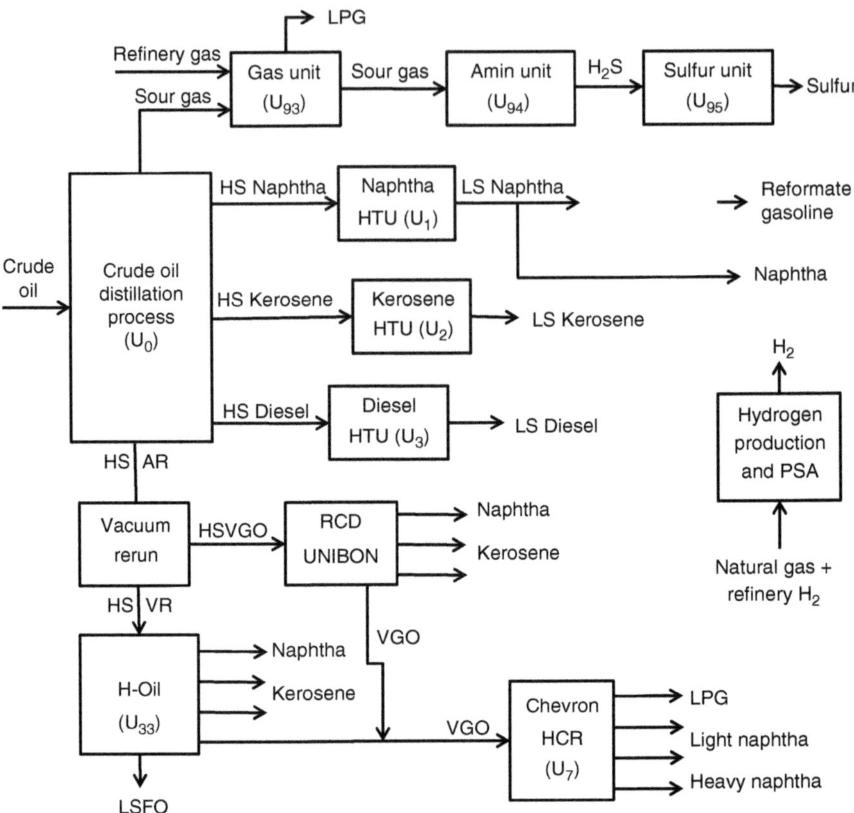

Figure 7.8 Simplified schematic representation of SHU refinery configuration, net refinery profit = US$ 16.76/bbl (39.90 cent/gal) of crude oil refined. Source: Albahri et al. (2018). Reproduced with permission of the American Chemical Society.

refinery and Mina Al-Ahmadi (MAA) refinery. We do this validation by assigning fixed values of 0 or 1 to the structural binary variables in the model to indicate unit existence or otherwise and compute the resulting material balances and profits for all intermediate and final product streams as summarized in Figures 7.8–7.10 for the respective refineries.

Figure 7.8 shows a schematic of SHU configuration that comprises a vacuum flasher followed by H-Oil process that performs a dual function of hydrocracking and HDT high-sulfur vacuum residue. The resulting low-sulfur VGO is then further hydrocracked.

Figure 7.9 shows MAA configuration that desulfurizes a high-sulfur AR stream through a similar dual function (as in H-Oil) of HDT and mild hydroconversion followed by vacuum flashing. The resulting low-sulfur vacuum residue is thermally cracked in a delayed coker, which is not part of the original scheme but added later to improve flexibility. The low-sulfur VGO is then sent to hydrocracker and FCC in equal proportion.

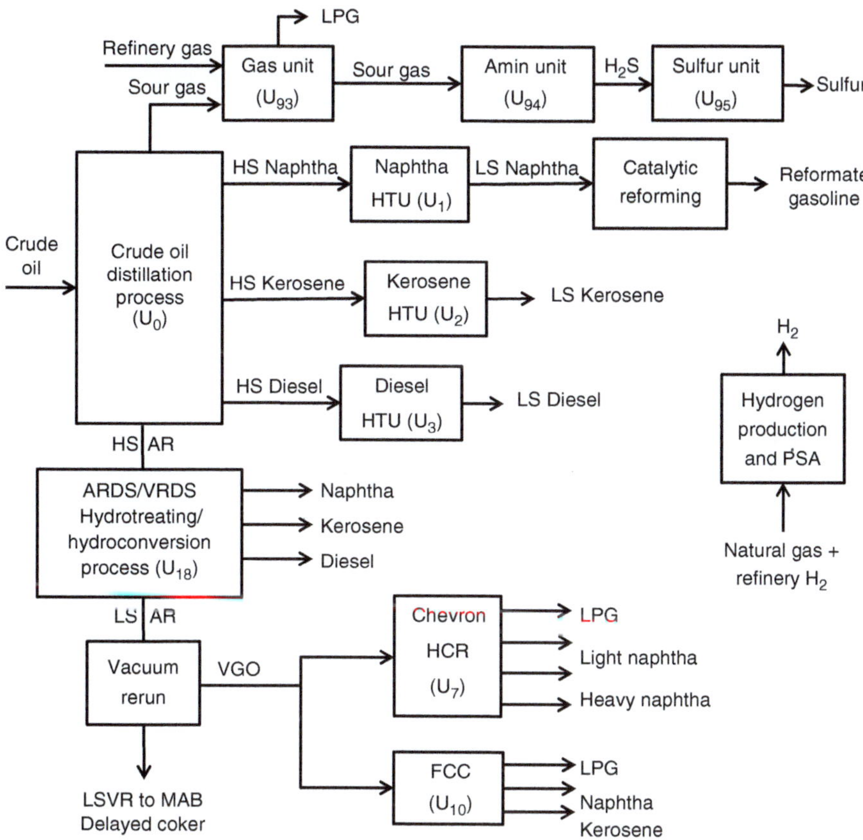

Figure 7.9 Simplified schematic representation of MAA refinery configuration, net refinery profit = US$ 17.27/bbl (41.10 cent/gal) of crude oil refined. Source: Albahri et al. (2018). Reproduced with permission of the American Chemical Society.

A simplified schematic of MAB configuration is shown in Figure 7.10 which is basically the same as MAA except that low-sulfur VGO is catalytically cracked by hydrocracking only; no FCC is required. The delayed coker here is part of the original scheme.

All three configurations involve routes that desulfurize the light products and middle distillates and produce hydrogen by catalytic steam reforming of natural gas. They also include refinery gas, amine, and sulfur recovery units. Both SHU and MAA also involve catalytic reforming of heavy straight-run naphtha; but not for MAB. The distribution of crude in various products in Table 7.11 shows that in all configurations, substantial light distillates are produced; only SHU produces fuel oil for sale. We also report the total investment and operating costs for the three operating refineries. The refinery configuration that is most profitable is MAB at US$ 18.00/bbl (42.86 cent/gal) net profit, followed by MAA, and SHU the least profitable.

The SHU configuration is analogous to that of VRDS scheme. As discussed previously, we expect SHU to be less profitable than those of the refineries based on the

Table 7.11 Refinery net product yields and associated economic evaluations for operating residue conversion refineries with 200,000 bbl/d (barrel per day) capacity.

		MAA Refinery		MAB Refinery		SHU Refinery	
Product		bbl/d	ton/d	bbl/d	ton/d	bbl/d	ton/d
Crude oil (Brent June 2014)		−200,000	−30,120	−200,000	−30,120	−200,000	−30,120
Natural gas (in cm^3/d, scfd)		−20,997,556	−436	−59,031,367	−1,225	−57,832,813	−1,200
Refinery gas (fuel oil equivalent, FOE)		9,679	5,083	7,433	5,051	7,195	4,759
Liquefied petroleum gas		12,060	1,050	5,132	447	5,002	435
Naphtha		49,790	5,553	40,530	4,521	42,727	4,766
Gasoline		76,730	8,558	94,828	10,577	86,431	9,640
Kerosene		27,820	3,542	27,820	3,542	27,820	3,542
Diesel		44,745	6,040	48,008	6,480	50,191	6,775
Heavy fuel oil		0	0	0	0	4,339	680
Petroleum sulfur			720		720		720
Petroleum coke (calcined)			1,439		1,350		0
Total capital cost		2,164,311,325		2,339,118,024		2,012,474,538	
Capital recovery (0%/d, 50 y, 365 d/y)		118,592		128,171		110,273	
Running cost (US$/d)		513,740		717,591		712,215	
Maintenance allowance, taxes, and insurance (US$/d)		177,889		192,256		165,409	
Total personnel (US$/d) (see note in Table 7.9)		550,000		550,000		550,000	
Natural gas cost (US$/d)		73,898		207,754		203,535	
Crude oil cost (US$/d)		21,600,000		21,600,000		21,600,000	
Total sales from products (US$/d)		26,487,287		26,996,081		26,694,191	
Refinery profit	US$/yr	1,260,406,368		1,314,113,044		1,223,757,041	
	US$/d	3,453,168		3,600,310		3,352,759	
	US$/bbl	17.27		18.00		16.76	
	cent/gal	41.10		42.90		39.90	
Internal rate of return (%)		6.2		6.1		6.6	

Source: Albahri et al. (2018). Reproduced with permission of the American Chemical Society.

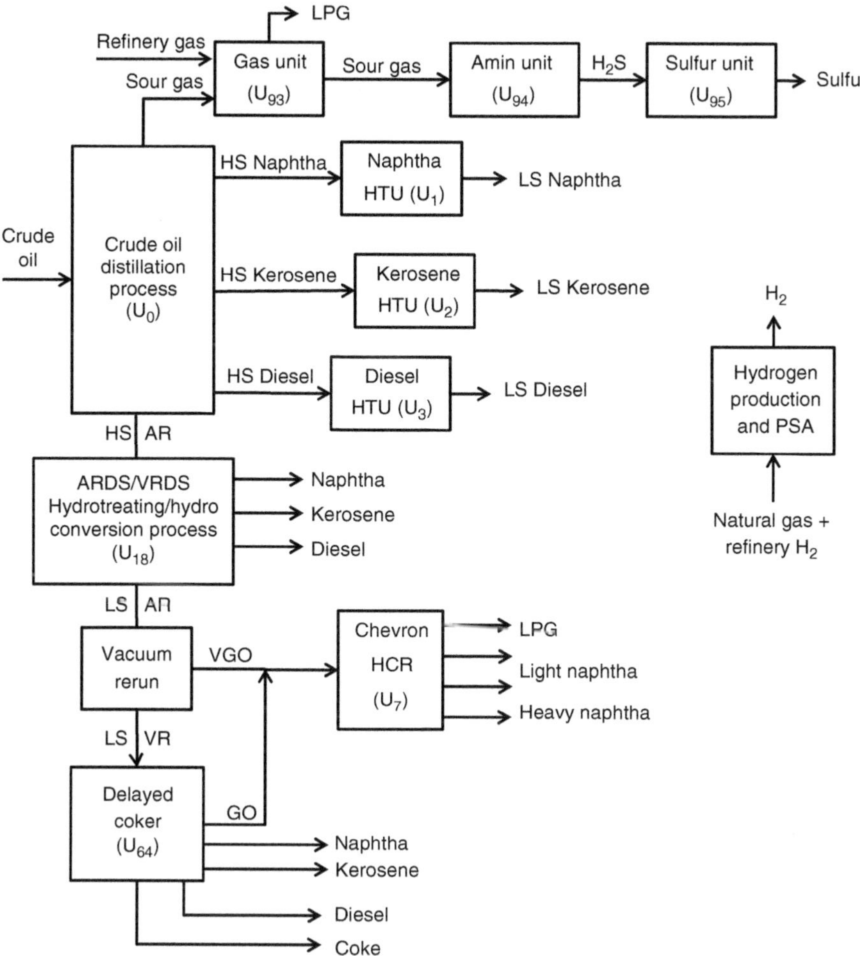

Figure 7.10 Simplified schematic representation of MAB refinery configuration, net refinery profit = US$ 18.00/bbl (42.90 cent/gal) of crude oil refined. Source: Albahri et al. (2018). Reproduced with permission of the American Chemical Society.

ARDS scheme, namely, the MAA and MAB refineries. MAB is the more modern of the three refineries and the most profitable as it uses hydrocracking and DCK; MAA is the next most profitable. However, all three configurations are less profitable than the RFCC-based optimal configuration that our model computes. Although SHU is based on VRDS scheme, it shows less profit than our model's computed profit (see Table 7.9), which results from optimally selecting the available units within the VRDS alternative by combining delayed coker and KRW gasifier. Such a combination achieves what is industrially called a complete-conversion refinery, which also provides supplementary hydrogen and is thus more profitable than the SHU configuration that uses H-Oil technology to process low-sulfur vacuum residue. On the same note, our model's ARDS configuration profit is higher than that of MAA and MAB refineries because it considers ARC process besides KRW gasifier.

Besides the complete conversion nature of DCK in ARDS scheme as compared to the incomplete conversion of H-Oil, another reason the ARDS scheme is preferred to that of VRDS is that the latter undergoes GO HDT, which increases the costs although it involves a smaller residue desulfurizer capacity. We also note a shift in modern refining trend from a VRDS-like configuration to that of ARDS (Axens 2014).

Our approach helps to give insight into complex refinery interactions that may not be apparent from reasoning without basing on a mathematical model or heuristics derived from experience. To further illustrate, fluctuating product prices seasonally can affect what constitutes an optimal configuration. For example, if fuel oil prices for grades number 2 and number 6 drop below a certain value, then catalytic hydrocracking becomes preferred than APC FCC process. In that case, hydrocracking low-sulfur GO in UOP/Unocal Unicracking gasoline mode is preferred than other methods. Therefore, it is essential to have representative economics data in refinery configuration studies.

7.2.2.3 Application Extension to Refinery Upgrade Studies

We can also use our model to perform studies to revamp and optimize existing refinery configurations by setting the binary structural variables to one for existing units and associating such variables with potential new processing technologies to be added for upgrading a refinery. For example, we carry out such a study of the foregoing MAB refinery and find that its net profit increases by US$ 0.71/bbl when we consider using KRW fluidized bed coal gasification process to convert the green coke product from delayed coker to hydrogen (instead of selling the coke as fuel).

7.2.2.4 Sensitivity Analysis

A sensitivity analysis of the model as reported in Figure 7.11 shows that the profit objective values for all three configurations increase rapidly with feed capacity until a throughput of 200,000 bpd before the increase becomes marginal. This trend implies at higher capacities, profit increases disproportionately to the investment. However, some process units are limited by physical considerations; consequently, we may require two or more parallel processing facilities or trains, thereby accruing more investment and reducing marginal profit increase, which is reported by Maiti (2001) as well.

7.2.3 Concluding Remarks

This work presents a technique to synthesize a petroleum refinery configuration using an aggregated network model to perform preliminary screening of the topology alternatives. The alternatives are developed based on three existing schemes for the desulfurization and cracking operations of crude oil mixtures (namely, ARDS, VRDS, and RFCC). We formulate an MILP that can be solved with standard computational resources to obtain an optimal solution that approximates a real-world refinery configuration. Consistent with industrial examples, our results indicate that an optimal crude oil processing scheme favors catalytic cracking of

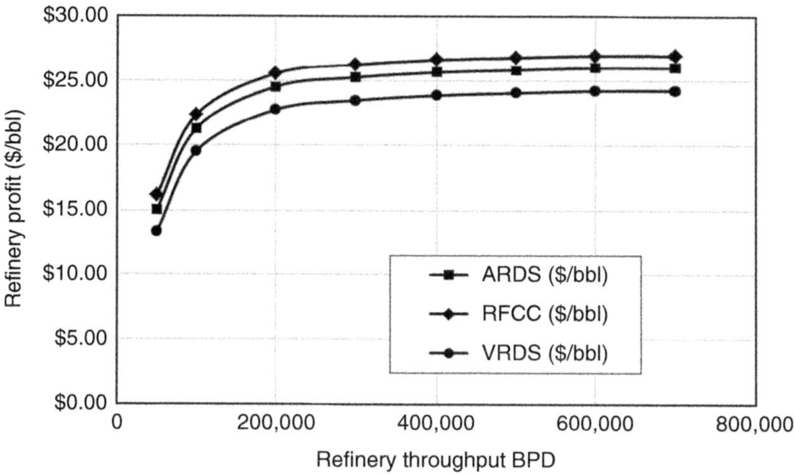

Figure 7.11 Sensitivity analysis of model results in terms of refinery profit versus throughput. Source: Albahri et al. (2018). Reproduced with permission of the American Chemical Society.

desulfurized AR as compared to physically separating the atmospheric and vacuum residues combined with thermal-, catalytic-, or hydrocracking. Our approach to incorporate logic propositions extensively within a mixed-integer optimization procedure allows us to explicitly consider complex refinery interactions as an alternative to using experience-based heuristics. We are extending this work to include more detailed representation through nonlinear process unit models, which gives rise to an MINLP formulation.

7.3 Industrial Case Study 3: Refinery Configuration for Naphtha Upgrading

Process synthesis or conceptual process design is concerned with the identification of the best flowsheet structure to perform a given task. Three major approaches are traditionally available in the literature to address this class of problem: (1) the heuristics method, notably the hierarchical decomposition of design decisions procedure; (2) the technique based on thermodynamic targets and physical insights as exemplified by pinch analysis; and (3) the algorithmic approach that utilizes optimization based on the construction of a superstructure that seeks to represent all feasible process flowsheets (Seider et al. 2009).

The intricate complexities associated with the process synthesis problem in general and the refinery design problem in specific necessitates the development and implementation of a systematic and automated approach that efficiently and rigorously integrates the elaborate interactions involving the design decision variables. This work aims to extend the superstructure optimization-based approach of using logical constraints (Raman and Grossmann 1991, 1992, 1993a,b) within a MILP to incorporate qualitative design knowledge based on engineering experience and heuristics in modeling the major process flows in a refinery. These constraints adopt discrete integer decision variables of the binary 0–1 type to model the existence of a refinery process unit and the associated stream piping interconnections (which are effectively pipelines) in a network structure, in which a value of one for a 0–1 variable designates that a unit is present in the optimal structure while the converse is true for a value of zero.

Our work serves to further substantiate that the use of 0–1 decision variables offers a more natural and powerful modeling approach compared to the conventional LP technique that employs only continuous decision variables. It also affords the convenience of representing fixed-cost charges in the objective function formulation.

7.3.1 Problem Statement

We consider the following process synthesis problem of superstructure optimization for the topology design of a refinery. Given the following data: (1) fixed production amounts of desired products, (2) the available process units and ranges of their capacities, and (3) cost of crude oil and cost structure for process units, we are to determine the optimal topology or configuration of the refinery in terms of the selection and sequencing of the streams as well as the operating levels as represented by the stream flowrates.

7.3.2 Propositional Logics and Logic Cuts in Process Synthesis Problems

Industrial Case Study 3 emphasizes the practical use of logical constraints on design and structural specifications for a refinery topology or configuration design to upgrade a naphtha stream. Logical constraints have been proven to be logic cuts that serve to reduce the computational expense of solving a MILP by tightening its linear relaxation and excluding fractional solutions without affecting the quality of the optimal solution (Hooker et al. 1994). They are algebraic linear inequalities or equalities formulated by using 0–1 binary variables to represent discrete decisions for the selection of alternative tasks corresponding to the process units as well as alternative states corresponding to the material streams.

7.3.3 Logical Constraints

7.3.3.1 General Formulation

Based on the superstructure of processing alternatives for naphtha exiting the ADU in Figure 7.12, we consider the following design specification: MIX–3 is selected if and only if LSRN–1 or LSRN–3 is produced. We contemplate the use of two logical relations and comment on some possible pitfalls.

First, using a combination of the logical "or" operator and the "equivalence" logic relation in the following form:

$$\left(Z_{\text{LSRN}-1} \vee Z_{\text{LSRN}-3}\right) \Leftrightarrow Y_{\text{MIX}-3} \tag{7.16}$$

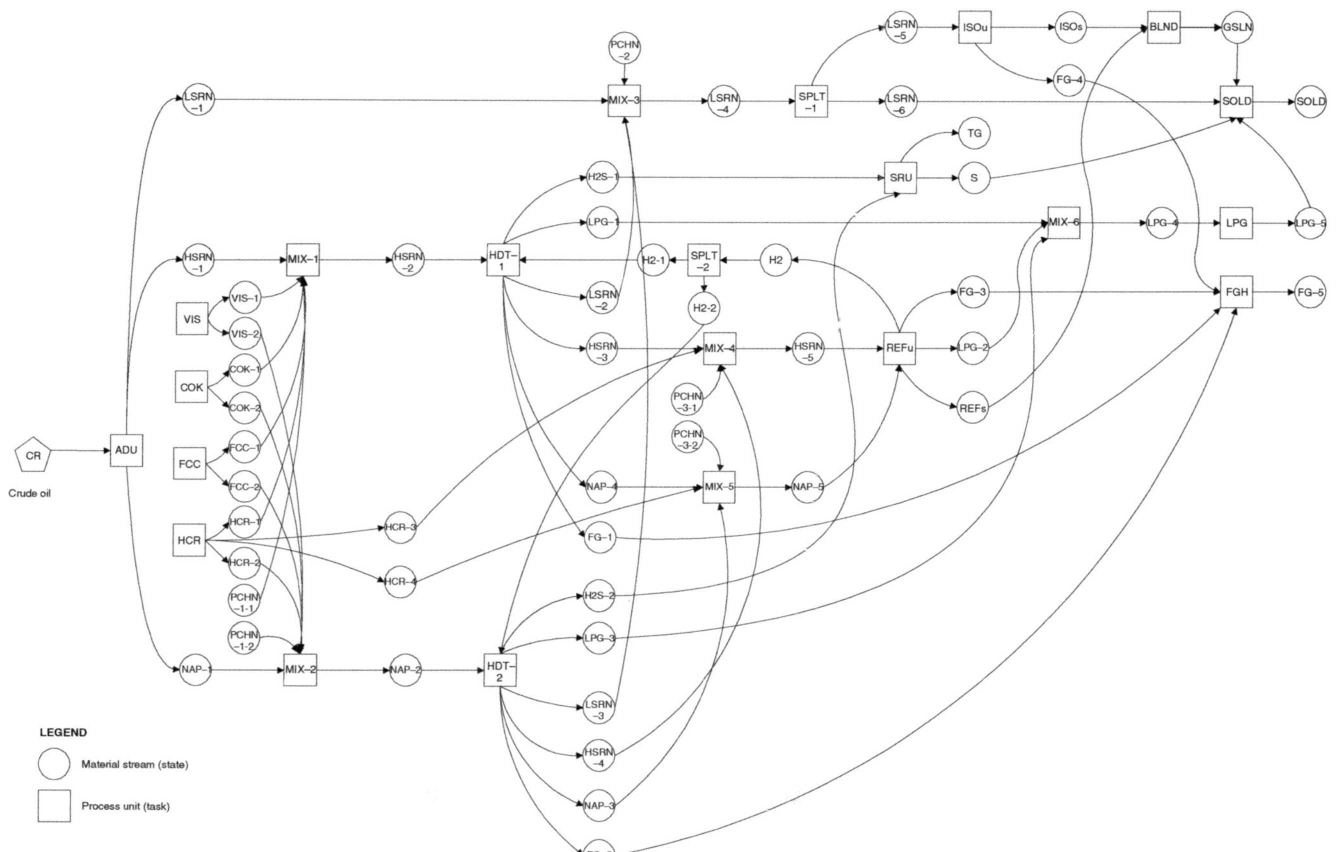

Figure 7.12 STN-based superstructure representation on case study of refinery topology for naphtha upgrading subsystem. Source: Asian Network for Scientific Information/CC BY 4.0.

This is equivalent to the following two logic propositions:

$$\begin{aligned}
&\left(Z_{\text{LSRN}-1} \vee Z_{\text{LSRN}-3}\right) \Rightarrow Y_{\text{MIX}-3}\\
&Y_{\text{MIX}-3} \Rightarrow \left(Z_{\text{LSRN}-1} \vee Z_{\text{LSRN}-3}\right)
\end{aligned} \tag{7.17}$$

Employing De Morgan's theorem yields the following:

$$\begin{aligned}
&\left(Z_{\text{LSRN}-1} \vee Z_{\text{LSRN}-3}\right) \Rightarrow Y_{\text{MIX}-3}\\
&\left(Z_{\text{LSRN}-1} \vee Z_{\text{LSRN}-3}\right) \vee Y_{\text{MIX}-3}\\
&\left(Z_{\text{LSRN}-1} \wedge Z_{\text{LSRN}-3}\right) \vee Y_{\text{MIX}-3}\\
&\underbrace{\left(Z_{\text{LSRN}-1} \vee Y_{\text{MIX}-3}\right)}_{1-z_{\text{LSRN}-1}+y_{\text{MIX}-3}\geq 1} \wedge \underbrace{\left(Z_{\text{LSRN}-3} \vee Y_{\text{MIX}-3}\right)}_{1-z_{\text{LSRN}-3}+y_{\text{MIX}-3}\geq 1}\\
&y_{\text{MIX}-3} \geq z_{\text{LSRN}-1} \ \text{and} \ y_{\text{MIX}-3} \geq z_{\text{LSRN}-3}
\end{aligned} \tag{7.18}$$

$$\begin{aligned}
&Y_{\text{MIX}-3} \Rightarrow \left(Z_{\text{LSRN}-1} \vee Z_{\text{LSRN}-3}\right)\\
&Y_{\text{MIX}-3} \vee \left(Z_{\text{LSRN}-1} \vee Z_{\text{LSRN}-3}\right)\\
&\left(1 - y_{\text{MIX}-3}\right) + z_{\text{LSRN}-1} + z_{\text{LSRN}-3} \geq 1\\
&z_{\text{LSRN}-1} + z_{\text{LSRN}-3} \geq y_{\text{MIX}-3}
\end{aligned} \tag{7.19}$$

thus, we obtain the following algebraic constraints:

$$\begin{aligned}
y_{\text{MIX}-3} &\geq z_{\text{LSRN}-1}\\
y_{\text{MIX}-3} &\geq z_{\text{LSRN}-3}\\
z_{\text{LSRN}-1} + z_{\text{LSRN}-3} &\geq y_{\text{MIX}-3}
\end{aligned} \tag{7.20}$$

However, the pitfall to using this formulation is that it allows the 0–1 variables to be satisfied for the case of $(z_{\text{LSRN-1}}, z_{\text{LSRN-3}}, y_{\text{MIX-3}}) = (1, 1, 1)$. This violates the physics of the problem stipulating that either LSRN–1 or LSRN–3 only is selected in the optimal configuration.

Consider now using the logical operator "exclusive or" as given by the following relation:

$$\left(Z_{\text{LSRN}-1} \veebar Z_{\text{LSRN}-3}\right) \Leftrightarrow Y_{\text{MIX}-3} \tag{7.21}$$

Translating this logic proposition into its equivalent algebraic constraints form, the proposition corresponds to:

$$z_{\text{LSRN}-1} + z_{\text{LSRN}-3} = y_{\text{MIX}-3} \tag{7.22}$$

$$z_{\text{LSRN}-1} + z_{\text{LSRN}-3} = 1 \tag{7.23}$$

However, there are three possible pitfalls in using this logical relation, which are all attributable to the logical constraint (7.23). First, this constraint compels either LSRN–1 stream or LSRN–3 stream to be selected even if there is no crude oil feed. Second, the two linear inequalities (7.22) and (7.23) enforce that $y_{\text{MIX-3}} = 1$, which mandates the MIX–3 unit to be selected under all circumstances; in other words, it requires MIX–3 to be a permanent feature of a refinery topology, which violates the physical problem. Third, this logic proposition is *not* satisfied for the case of ($z_{\text{LSRN-1}}$, $z_{\text{LSRN-3}}$, $y_{\text{MIX-3}}$) = (0, 0, 0), which is the hypothetical case of no crude oil feed is available. Thus, the constraints given by Eq. (7.20) best enforce the design specification that "MIX 3 is selected if and only if LSRN–1 or LSRN–3 is produced."

In our computational experiments, it is noteworthy to highlight the following frequently encountered form of logic proposition in developing logical constraints on design specifications and structural specifications for synthesis problems. The logic form is generally given as:

$$\bigvee_{k=1,2,\ldots,M} Y_{u,k} \Leftrightarrow Y_u, \quad \forall u \in U = \{1, 2, \ldots, N\} \tag{7.24}$$

which is equivalent to:

$$\left(\bigvee_{k=1,2,\ldots,M} Y_{u,k} \Rightarrow Y_u\right) \wedge \left(Y_u \Rightarrow \bigvee_{k=1,2,\ldots,M} Y_{u,k}\right), \quad \forall u \in U \tag{7.25}$$

Transforming these logic propositions into inequalities yields the following algebraic constraints:

$$\begin{aligned} \bigvee_{k=1,2,\ldots,M} Y_{u,k} \Rightarrow Y_u \quad & \forall u \in U \\ Y_{u,k} \Rightarrow Y_u \quad & \forall u \in U, k = 1, 2, \ldots, M \\ Y_{u,k} \vee Y_u \quad & \forall u \in U, k = 1, 2, \ldots, M \\ \left(1 - y_{u,k}\right) + y_u \geq 1 \quad & \forall u \in U, k = 1, 2, \ldots, M \\ y_u - y_{u,k} \geq 0 \quad & \forall u \in U, k = 1, 2, \ldots, M \end{aligned} \tag{7.26}$$

as well as for the logic proposition in the opposing direction:

$$\begin{aligned} & Y_u \Rightarrow \bigvee_{k=1,2,\ldots,M} Y_{u,k} \quad \forall u \in U \\ & Y_u \vee \left(\bigvee_{k=1,2,\ldots,M} Y_{u,k}\right) \quad \forall u \in U \\ & \left(1 - y_u\right) + \sum_{k=1}^{M} y_{u,k} \geq 1 \quad \forall u \in U \\ & \sum_{k=1}^{M} y_{u,k} - y_u \geq 0 \quad \forall u \in U \end{aligned} \tag{7.27}$$

The MILP model formulation is summarized as follows:

$$\begin{aligned} \min \;\; & Z = \sum c_u + f(x) + d^T y \\ \text{s.t.} \;\; & g_u(x) \leq 0 \\ & f_u \leq M_u y_u \\ & y_u - y_{u,k} \geq 0 \quad \forall i \in I, k = 1, 2, \ldots, M \\ & \sum_{k=1}^{M} y_{u,k} - y_u \geq 0 \quad \forall i \in I \\ & x \in \Re^n, \;\; y_u \in \{0, 1\}, \;\; c_u \geq 0, \;\; c \in \Re^m \end{aligned} \tag{7.28}$$

7.3.3.2 Logical Constraints on Processing Alternatives of Naphtha for Petroleum Refineries

This section summarizes the rest of the complete set of logical statements and their associated logic propositions for the subsystem of naphtha produced from ADU:

ADU is selected if and only if HDT–1 or HDT–2 is selected:

$$Y_{\text{ADU}} \Leftrightarrow \left(Y_{\text{HDT}-1} \underline{\vee} Y_{\text{HDT}-2}\right) \tag{7.29}$$

HDT–1 or HDT–2 is selected if and only if SRU is selected:

$$\left(Z_{\text{H2S}-1} \underline{\vee} Z_{\text{H2S}-2}\right) \Leftrightarrow Y_{\text{SRU}} \tag{7.30}$$

HSRN–3 or HSRN–4 is selected if and only if MIX–4 is selected:

$$\left(Z_{\text{HSRN}-3} \vee Z_{\text{HSRN}-4}\right) \Leftrightarrow Y_{\text{MIX}-4} \tag{7.31}$$

NAP–3 or NAP–4 is selected if and only if MIX–5 is selected:

$$\left(Z_{\text{NAP}-3} \vee Z_{\text{NAP}-4}\right) \Leftrightarrow Y_{\text{MIX}-5} \tag{7.32}$$

HDT–1 or HDT–2 is selected if and only if MIX–3 and MIX–4, or MIX–3 and MIX–5, or MIX–5 is selected:

$$\left(Y_{\text{HDT}-1} \vee Y_{\text{HDT}-2}\right) \Leftrightarrow \left(Y_{\text{MIX}-3} \wedge Y_{\text{MIX}-4}\right) \vee \left(Y_{\text{MIX}-3} \wedge Y_{\text{MIX}-5}\right) \vee Y_{\text{MIX}-5} \tag{7.33}$$

HSRN–5 or NAP–5 is selected if and only if REFu is selected:

$$\left(Z_{\text{HSRN}-5} \underline{\vee} Z_{\text{NAP}-5}\right) \Leftrightarrow Y_{\text{REFu}} \tag{7.34}$$

HDT–1 or HDT–2 is selected if and only if LPG is selected:

$$\left(Y_{\text{HDT}-1} \vee Y_{\text{HDT}-2}\right) \Leftrightarrow Y_{\text{LPG}} \tag{7.35}$$

FG–1 or FG–2 or FG–3 or FG–4 is selected if and only if FGH is selected:

$$\left(Z_{\text{FG}-1} \vee Z_{\text{FG}-2} \vee Z_{\text{FG}-3} \vee Z_{\text{FG}-4}\right) \Leftrightarrow Y_{\text{FGH}} \tag{7.36}$$

ISO is selected if and only if HDT–1 is selected:

$$\left(Z_{\text{ISO}} \vee Z_{\text{REF}}\right) \Leftrightarrow Y_{\text{BLND}} \tag{7.37}$$

7.3.4 Computational Experience

Computational experiments and numerical studies on the proposed mixed-integer optimization framework of MILP and GDP models for the flowsheet superstructure optimization problem are implemented using the GAMS modeling language platform. Two design scenarios as distinguished by the API gravity of the crude oil charge to the ADU are considered in the computational experiments conducted, namely, for light crude oil mixture as characterized by API > 33° and heavy crude oil mixture as characterized by API ≤ 33°. Both scenarios are developed based on conventional distillation column design and refinery economics, in which products from ADU are either separated into light straight run naphtha (LSRN–1) and heavy straight

Table 7.12 Optimal objective function values.

	Light crude oil processing		Heavy crude oil processing	
Model type	**MILP**	**GDP**	**MILP**	**GDP**
Objective value (in million MYR per year)	2744	2465	2743	2453

run naphtha (HSRN–1) streams or as an undifferentiated class of naphtha (NAP–1), which is also sometimes called wild naphtha in the industry.

The optimal refinery topology or configuration generated by the MILP and the GDP is, as expected, identical for both design scenarios. As depicted in Figures 7.13 and 7.14, the optimal solutions of both models select an identical processing route comprising the same process units and material streams and their interconnections but at different processing levels. It is noteworthy that the optimal topology generated by each formulation agrees reasonably well with the topology of real-world existing refineries reported in standard references on the refining industry (Maples 2000b; Gary et al. 2007; Meyers 2016; Hydrocarbon Processing 2019b). A closer inspection of the aforementioned figures reveals that it is economically optimal to build a reformer and a hydrotreater within a single site based on the particular economics considered in the numerical example.

The optimal objective function value of total investment cost (which accounts for a summation of the fixed capital costs, variable operating costs, and the raw material costs of purchasing the required crude oil slate) is validated by comparing the values tabulated in Table 7.12. Optimal objective function values against industrial data available in the open literature. For instance, an annualized total investment cost reported by an Indian refinery located in Mumbai is estimated to be approximately RM2700 million, which by inspection is of relatively reasonable accuracy with our results (Hydrocarbon Processing 2019a). It is noteworthy that in this respect, the GDP model offers a more cost-effective solution by registering a lower total investment cost.

The associated model sizes and computational statistics are reported in Table 7.13. As observed from the table, a GDP formulation typically offers a smaller model size relative to its MILP equivalent hence, it is likely to be more amenable for implementing extensions to this work that encompasses design features required for a more complex refinery.

7.3.5 Concluding Remarks

This case study illustrates improved computational performance through incorporating engineering insights in process synthesis problems by using logical constraints on certain design and structural specifications. This is accomplished within a conventional MILP framework that has the advantage of considering the effects of all relevant constraints simultaneously, thus affording a global perspective

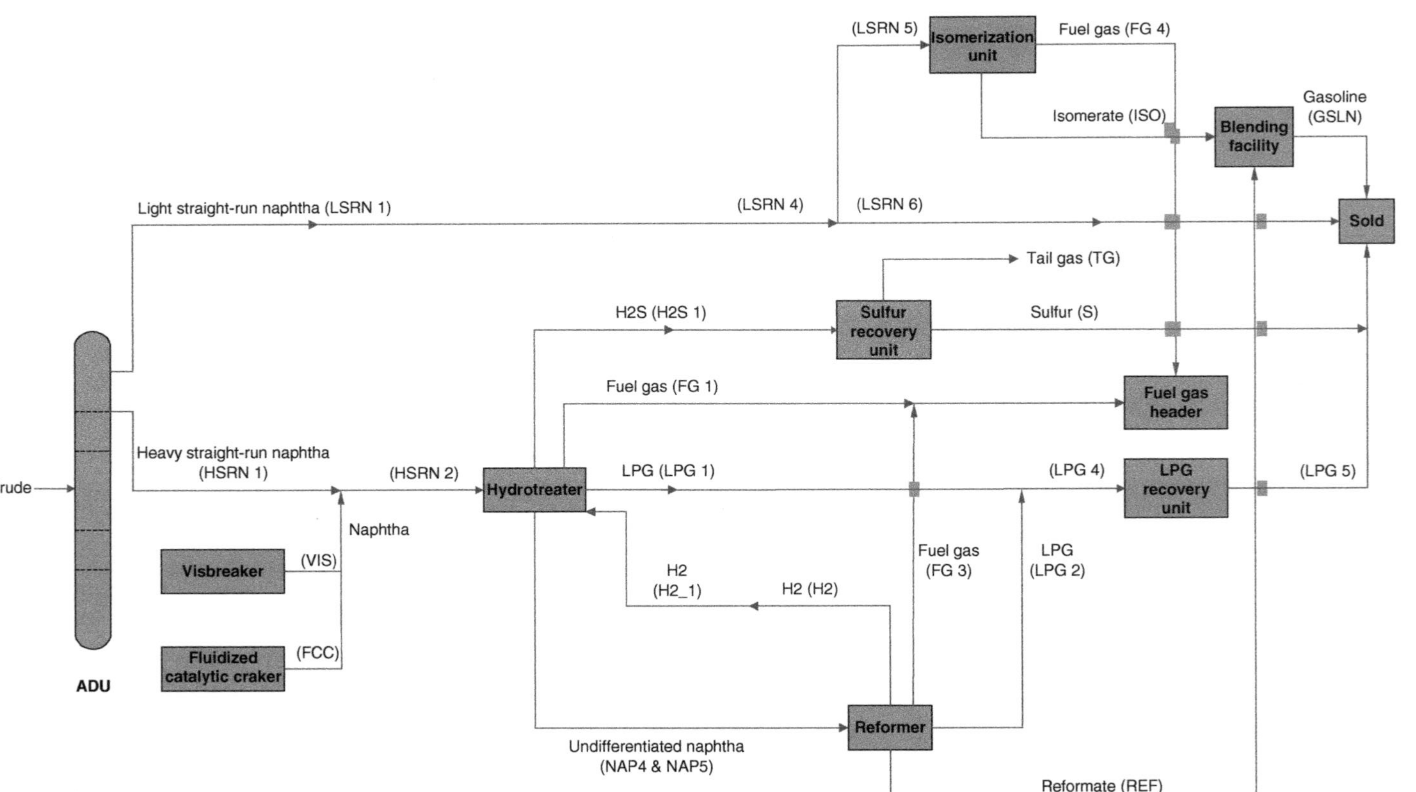

Figure 7.13 Optimal refinery topology of the naphtha processing subsystem for light crude oil charge.

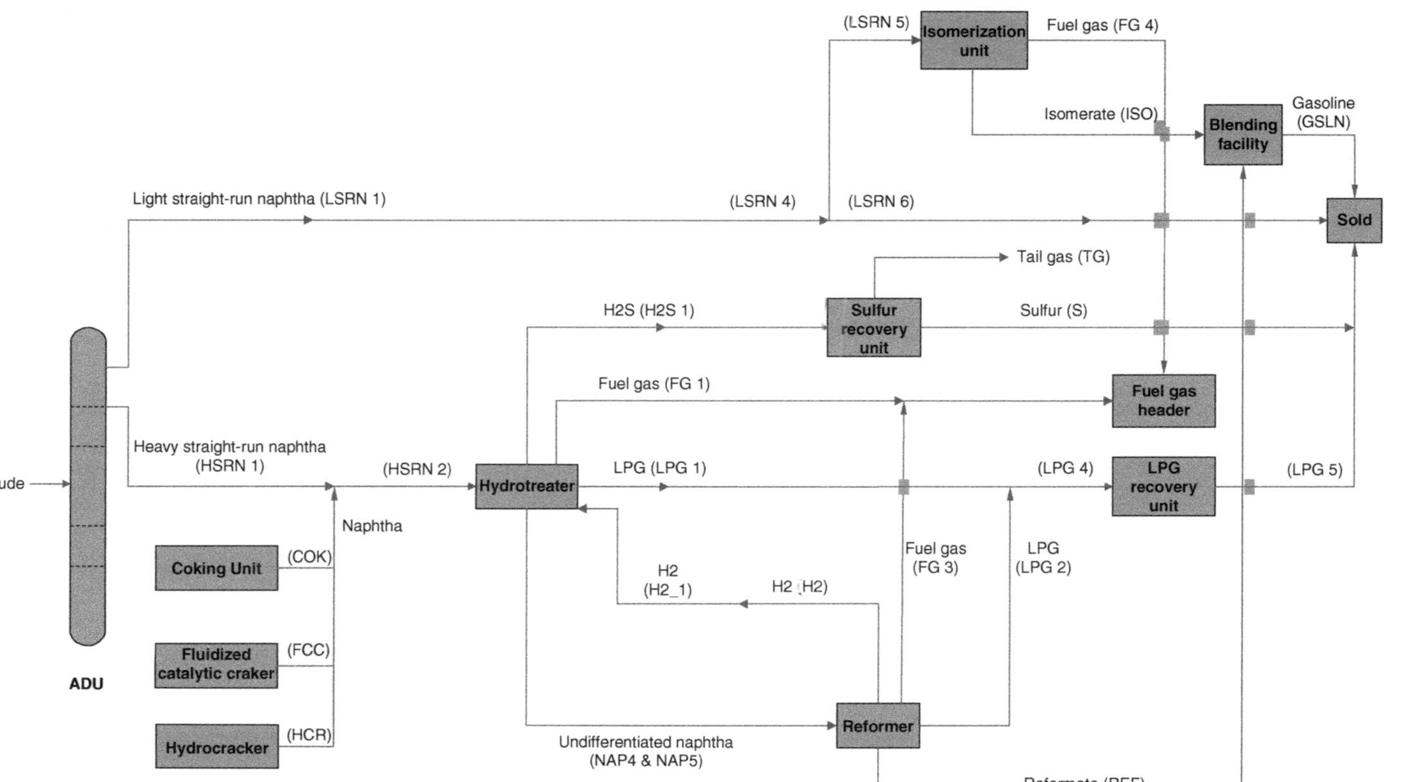

Figure 7.14 Optimal refinery topology of the naphtha processing subsystem for heavy crude oil charge.

Table 7.13 Model sizes and computational statistics.

Model type	**MILP**	**GDP**
Solver	**GAMS/CPLEX**	**GAMS/LogMIP**
Number of constraints	336	195
Number of continuous variables	77	95
Number of binary variables	80	22
Number of iterations	26	23
CPU time (s)	0.031	0.030

to the model. However, since the typical algorithm for MILP requires solving an LP subproblem at each node of the search tree, all the constraints must be linear equalities or inequalities. This imposes a restriction on the expressiveness of MILP as a modeling language as some problems may require a very large number of variables and constraints, which in turn gives rise to our attempt to adopt the alternative modeling framework of GDP.

7.4 Chapter Summary

This chapter presents several applications of the foregoing optimization modeling framework introduced in earlier chapters. Four case studies on configuration design are illustrated on problems of industrial size and significance for the petroleum refining industry. Each of the applied examples emphasizes differing aspects of importance to this process synthesis problem type.

References

Albahri, T.A., Al-Sharrah, G., Khor, C.S., and Elkamel, A. (2019a). Grassroots petroleum refinery configuration for heavy oil processing. *Petroleum Science and Technology* 37 (3): 275–281.

Albahri, T.A., Khor, C.S., Elsholkami, M., and Elkamel, A. (2018). Optimal design of petroleum refinery configuration using a model-based mixed-integer programming approach with practical approximation. *Industrial & Engineering Chemistry Research* 57 (22): 7555–7565.

Albahri, T.A., Khor, C.S., Elsholkami, M., and Elkamel, A. (2019b). A mixed integer nonlinear programming approach for petroleum refinery topology optimisation. *Chemical Engineering Research and Design* 143: 24–35.

Ancheyta, J. and Speight, J.G. (2007). *Hydroprocessing of Heavy Oils and Residua*. Boca Raton, FL: CRC Press.

Axens (2014). Axens' Technologies, Products and Services. www.axens.net (accessed 1 November 2021).

Biegler, L.T., Grossmann, I.E., and Westerberg, A.W. (1997). *Systematic Methods of Chemical Process Design*. New Jersey: Prentice Hall.

Caballero, J.A. and Grossmann, I.E. (1999). Aggregated models for integrated distillation systems. *Industrial & Engineering Chemistry Research* 38 (6): 2330–2344.

Castillo Castillo, P., Castro, P.M., and Mahalec, V. (2017). Global optimization algorithm for large-scale refinery planning models with bilinear terms. *Industrial & Engineering Chemistry Research* 56 (2): 530–548.

Elshout, R.V., Bailey, J., Brown, L., and Nick, P. (2018). Upgrading the bottom of the barrel. *Hydrocarbon Processing* (March).

Fibre2Fashion Pvt Ltd (2017). Naphtha price comparison report, naphtha market trend, naphtha industry price trend, market watch report – Fibre2Fashion (cited July 12 2017). http://www.fibre2fashion.com/market-intelligence/textile-market-watch/naphtha-price-trends-industry-reports/22/ (accessed 1 November 2021).

FillLPG.co.uk. (2017). FillLPG.co.uk – Your LPG filling station map (cited July 11 2017). https://www.filllpg.co.uk/ (accessed 1 November 2021).

Fitzgibbon, T., Ding, C., and Szabat, P. (2018). Diesel demand: still growing globally despite Dieselgate. McKinsey (cited 15 January 2019). https://www.mckinsey.com/industries/oil-and-gas/our-insights/petroleum-blog/diesel-demand-still-growing-globally-despite-dieselgate (accessed 1 November 2021).

Frecon, J., Le-Bars, D., and Rault, J. (2019). Flexible upgrading of heavy feedstocks. *Petroleum Technology Quarterly* 24 (1): 31–39.

Frontline Systems. (2018). FrontlineSolvers. Frontline Systems, Inc. (cited 6 February 2018). https://www.solver.com/ (accessed 1 November 2021).

Gary, J.H. and Handwerk, G.E. (1994). *Petroleum Refining: Technology and Economics*, 3e, 204. New York: Marcel Dekker.

Gary, J.H., Handwerk, G.E., and Kaiser, M.J. (2007). *Petroleum Refining: Technology and Economics*, 5e. New York: Marcel Dekker.

Gray, M.R. (1994). *Upgrading Petroleum Residues and Heavy Oils*. New York: Marcel Dekker.

Grossmann, I.E. (2005). Advances in logic-based optimization approaches to process integration and supply chain management. *Chemical Engineering Trends and Developments* 299–322.

Hartmann, J.C. (1999). Interpreting LP outputs. *Hydrocarbon Processing* 78: 64–68.

Hooker, J.N., Yan, H., Grossmann, I.E., and Raman, R. (1994). Logic cuts for processing networks with fixed charges. *Computers & Operations Research* 21 (3): 265–279.

Hydrocarbon Processing (2011). *Refining Processes 2011 Handbook*. Houston, TX: Gulf Publishing.

Hydrocarbon Processing. (2019a). HPI Construction Boxscore.

Hydrocarbon Processing (2019b). *Refining Processes 2019 Handbook*. Houston, TX: Gulf Publishing.

Ibrahim, D., Jobson, M., Li, J., and Guillén-Gosálbez, G. (2018). Optimization-based design of crude oil distillation units using surrogate column models and a support vector machine. *Chemical Engineering Research and Design* 134: 212–225.

IndexMundi. (2017). Crude oil (petroleum) – monthly price – commodity prices – price charts, data, and news – IndexMundi (cited 11 July 2017). http://www.indexmundi.com/commodities/?commodity=crude-oil (accessed 1 November 2021).

Kamiya, Y. (1991). *Heavy Oil Processing Handbook*. Tokyo, Japan: Research Association for Residual Oil Processing (RAROP).

Khor, C. (2019a). High-octane gasoline production from catalytic naphtha reforming. In: *Petroleum Chemicals* (ed. M. Zoveidavianpoor), 1–15. London, UK: InTechOpen.

Khor, C. (2019b). A model-based investment assessment for heavy oil processing in the petroleum refining industry. In: *Processing of Heavy Crude Oils – Challenges and Opportunities* (ed. M. Ramasamy), 1–15. London, UK: InTechOpen.

Khor, C.S. (2019c). A model-based investment assessment for heavy oil processing in the petroleum refining industry. In: *Processing of Heavy Crude Oils – Challenges and Opportunities* (ed. R.M. Gounder). IntechOpen.

Khor, C.S. and Elkamel, A. (2010). Superstructure optimization for oil refinery design. *Petroleum Science and Technology* 28 (14): 1457–1465.

Khor, C.S. and Elkamel, A. (2016). Optimal synthesis of a heat-integrated petroleum refinery configuration. *The Canadian Journal of Chemical Engineering* 94 (10): 1939–1946.

Khor, C.S., Loh, C.Y., and Elkamel, A. (2008). A logic-based mixed-integer superstructure optimization approach for the optimal design of petroleum refinery topology with environmental considerations. In: *5th International Conference on Foundations of Computer-Aided Process Operations (FOCAPO)* (ed. M.H. Bassett, M. Ierapetritou, and E. Pistikopoulos), 221–224. Massachusetts, USA: Cambridge.

Khor, C.S. and Varvarezos, D. (2017). Petroleum refinery optimization. *Optimization and Engineering* 18 (4): 943–989.

Leiras, A., Ribas, G., Hamacher, S., and Elkamel, A. (2013). Tactical and operational planning of multirefinery networks under uncertainty: an iterative integration approach. *Industrial & Engineering Chemistry Research* 52 (25): 8507–8517.

Maiti, S., Eberhardt, J., Kundu, S. et al. (2001). How to efficiently plan a grassroots refinery. *Hydrocarbon Processing* 80: 43–49.

Maples, R.E. (2000a). *Petroleum Refinery Process Economics*, 2e, 264. Oklahoma: Pennwell.

Maples, R.E. (2000b). *Petroleum Refinery Process Economics*, 2e. Oklahoma: Pennwell.

Messick, D. (2012). World Sulphur Outlook (cited 17 January 2019). http://www.firt.org/sites/default/files/donmessick_sulphur_outlook.pdf (accessed 1 November 2021).

Meyers, R.A. (2016). *Handbook of Petroleum Refining Processes*, 4e. McGraw Hill.

Petrochemical Update. (2017). *Global naphtha surplus forecast to grow; Rising oil product exports split US-Mexico trade balance*. FCBI Energy Ltd (cited 21 January 2019). http://analysis.petchem-update.com/engineering-and-construction/global-naphtha-surplus-forecast-grow-rising-oil-product-exports-split (accessed 1 November 2021).

Pinto, J.M., Joly, M., and Moro, L.F.L. (2000). Planning and scheduling models for refinery operations. *Computers & Chemical Engineering* 24 (9): 2259–2276.

QPSPP. (2017). Qatar petroleum for the sale of Petroleum Products Company Limited (QPSPP) 2017 (cited 11 July 2017). https://www.qp.com.qa/en/marketing/Pages/QPSPP.aspx (accessed 1 November 2021).

Rabiei, Z. (2012). Hydrogen management in refineries. *Petroleum & Coal* 54 (4): 357–368.

Raman, R. and Grossmann, I.E. (1991). Relation between MILP modelling and logical inference for chemical process synthesis. *Computers & Chemical Engineering* 15 (2): 73–84.

Raman, R. and Grossmann, I.E. (1992). Integration of logic and heuristic knowledge in MINLP optimization for process synthesis. *Computers & Chemical Engineering* 16 (3): 155–171.

Raman, R. and Grossmann, I.E. (1993a). Relation between MILP modeling and logical inference for chemical process synthesis. *Computers & Chemical Engineering* 15: 73–84.

Raman, R. and Grossmann, I.E. (1993b). Symbolic integration of logic in mixed integer linear programming techniques for process synthesis. *Computers & Chemical Engineering* 17: 909–927.

Savant, A. (2018). Petroleum coke market report overview, global price trend, increasing demand, updated leading countries, industry analysis and future forecasts 2018–2025. Reuters (cited 15 January 2019). https://www.reuters.com/brandfeatures/venture-capital/article?id=61034 (accessed 1 November 2021).

Seider, W.D., Seader, J.D., Lewin, D.R., and Widagdo, S. (2009). *Product and Process Design Principles*, 3e. New York: Wiley.

Speight, J.G. (2011a). Chapter 2 – Refining processes. In: *The Refinery of the Future* (ed. J.G. Speight), 57. Boston: William Andrew Publishing.

Speight, J.G. (2011b). Chapter 6 – Catalytic cracking. In: *The Refinery of the Future* (ed. J.G. Speight), 181–208. Boston: William Andrew Publishing.

Speight, J.G. (2017). *Handbook of Petroleum Refining Boca Raton*. Florida: CRC Press.

Speight, J.G. and Özüm, B. (2002). Petroleum refining processes. In: *Chemical Industries* (ed. H. Heinemann). New York, Basel: Marcel Dekker.

SunSirs-China Commodity Data Group. (2017). Bulk commodity index, spot price: weekly, spot price: monthly (cited 11 July 2017). http://www.sunsirs.com/uk (accessed 1 November 2021).

Temizer, M. (2017). *Global LPG demand to rise sharply in 10 years*. Anadolu Agency (cited 20 May 2019). https://www.aa.com.tr/en/energy/natural-gas/global-lpg-demand-to-rise-sharply-in-10-years/1334 (accessed 1 November 2021).

U.S. Energy Information Administration. (2017). EIA independent statistics and analysis 2017 (cited 11 July 2017). http://www.eia.gov/dnav/pet/pet_pri_refoth_dcu_nus_m.htm (accessed 1 November 2021).

UOP (2010). Driving optimization and profitability through technology innovation. *Hydrocarbon Processing, Special Supplement* 6.

Zhang, J., Zhu, X.X., and Towler, G.P. (2001). A simultaneous optimization strategy for overall integration in refinery planning. *Industrial and Engineering Chemistry Research* 40 (12): 2640–2653.

Zhang, N. and Zhu, X.X. (2000). A novel modelling and decomposition strategy for overall refinery optimisation. *Computers & Chemical Engineering* 24 (2): 1543–1548.

Zhang, N. and Zhu, X.X. (2006). Novel modelling and decomposition strategy for total site optimisation. *Computers & Chemical Engineering* 30 (5): 765–777.

8

Industrial Case Studies with Environmental-Centric Techno-Commercial Considerations

This chapter discusses two case studies to illustrate refinery design configuration with environmental-centric techno-commercial considerations.

8.1 Industrial Case Study 1: Refinery Configuration with Environmental Considerations

This case study posits optimal petroleum refinery topology as a conceptual process synthesis problem. We begin by developing a sufficiently rich STN-based superstructure representation that encompasses all possible topology alternatives of a conventional refinery. Subsequently, a biobjective mixed-integer linear program (MILP) of profit maximization and environmental impacts minimization is formulated according to the constructed superstructure. Then, based on a given set of fixed amounts of desired products, the model is solved to generate an optimal topology. The proposed optimization framework also incorporates principles from life cycle analysis (LCA) to account for potential environmental impacts. A numerical example is illustrated to implement the modeling approach.

8.1.1 Background

Process synthesis or conceptual process design is concerned with the identification of the best flowsheet structure to perform a given task. Three major approaches have been reported in the literature to address these problems: (i) the heuristics method, notably the hierarchical decomposition of design decisions procedure (Douglas 1985); (ii) the technique based on thermodynamic targets and physical insights, as exemplified by pinch analysis (Linnhoff 1993); and (iii) the algorithmic approach, which utilizes optimization or mathematical programming, based on the construction of a superstructure that seeks to represent all feasible process flowsheets (Grossmann 2002). The complexity associated with synthesis problems in general and petroleum refinery design, in particular, necessitates the development and implementation of a systematic and automated approach to efficiently and rigorously consider the elaborate interactions and trade-offs among the design variables. In this regard, powerful formal optimization strategies potentially offer promising tools to undertake the task.

Hence, this section seeks to investigate the viability of employing a logic-based mixed-integer superstructure optimization approach in the preliminary screening of design alternatives for a grassroots refinery using an aggregated model with a high level of abstraction. Logical constraints based on propositional logic inferences are employed to model qualitative design knowledge based on engineering experience and heuristics, in order to stipulate the design specifications and structural specifications on the interconnectivity relationships among the process units and the material streams of a refinery. This associated model is formulated in accordance with the approach advocated by Raman and Grossmann (1991, 1992, 1993, 1994).

8.1.2 Problem Statement

We consider the following process synthesis problem of superstructure optimization of a refinery topological design. Given the following information: (i) fixed production amounts in terms of a set of desired products; (ii) available process units and the ranges of their capacities; (iii) cost of crude oil (feedstock) and capital expenditure for process units; determine: (i) the optimal topology or configuration of the refinery in terms of the selection and sequencing of the streams and the units; and (ii) the flowrates of the components in each stream.

8.1.3 Model Formulation

8.1.3.1 Superstructure Representation

The first step involves constructing a suitable superstructure representation for the problem, which leads to an appropriate mathematical formulation. Figure 8.1 shows an STN-based superstructure representation that is sufficiently rich to embed all feasible alternative topologies for a refinery. To facilitate the superstructure development, a typical refinery network is considered to be decomposed into four processing pools: (i) naphtha exiting the atmospheric distillation unit (ADU); (ii) reduced crude from the ADU and the vacuum distillation unit (VDU); (iii) vacuum gas oil (VGO) from VDU; and (iv) heavy oil processing and upgrading.

At this point, the next step is to decide on a general solution strategy. Since the model developed is largely linear, a simultaneous optimization strategy is adopted instead of a sequential approach.

8.1.3.2 Material Balance Constraints

The overall and component material balances in the form of input–output of flowrates are given by the following general linear relation:

$$Ax = 0 \tag{8.1}$$

where A = matrix of linear constants or fixed yields as obtained from the literature (Kamiya 1991; Gary et al. 2007) and x = decision variables vector of material flowrates. The other sets of material balances are for the modeling units of mixers

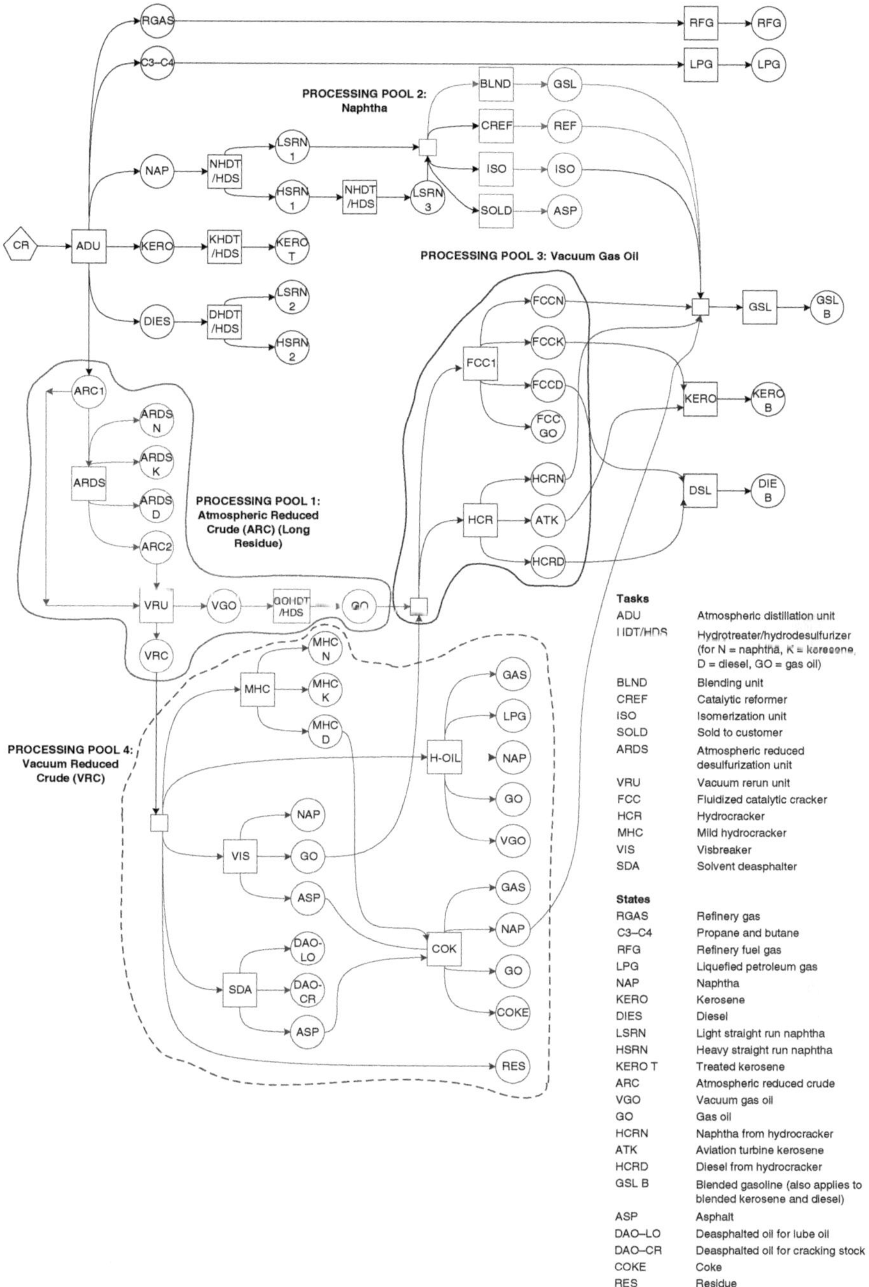

Figure 8.1 Superstructure for a petroleum refinery topology based on state–task network (STN) representation.

and splitters, which model the interconnections of the process units and are, therefore, convenient key modules for developing the material balances. As well, constraints specifying the variables' lower and upper bounds on the material flows and the upper bounds on production requirements are included.

8.1.3.3 Logical Constraints

In this work, we employ logical constraints according to the approach advocated by Raman and Grossmann (1991, 1993) to incorporate qualitative design knowledge based on engineering experience and heuristics in modeling the major process flows in a refinery. We accomplish this by utilizing logic propositions and integer variables. The logical constraints play the following two roles:

1. to enforce design specifications on the selection of the units and the streams linking the units – these specifications mainly describe the inherent characteristics of the units based primarily on engineering knowledge and past design experience; and
2. to enforce structural specifications that stipulate the interconnectivity relationships among the nodes of units and streams in the network – these relationships describe the sequence in which the streams are linking the units.

In addition, we need to impose the relations between the continuous variables representing the flowrates of the streams and the discrete 0–1 variables to ensure that the nonexistence of a unit results in zero input flows, as well as the output flows (by extension from the material balances). These relationships can be expressed by logical constraints using big-M formulation, which are also termed as switching constraints.

8.1.3.4 Logic Propositions

The major process flows in a refinery network are discussed in this section, emphasizing the logical constraints formulation as described earlier. These constraints model the related qualitative information by using features afforded by propositional logics and binary 0–1 variables. A binary variable with the value of one indicates that a processing unit or material stream is selected in an optimal topology solution; a value of zero indicates otherwise.

8.1.3.4.1 Processing Pool 1: Alternatives for Atmospheric Reduced Crude (ARC)

The crude oil from the storage tank is heated in a furnace and then charged to an atmospheric crude distillation unit (ADU), which is a mainstay feature of an oil refining scheme as the primary fractionation function of the crude oil according to different boiling point ranges. ADU separates the crudes into butanes and lighter wet gases, unstabilized light naphtha, heavy naphtha, kerosene, atmospheric gas oil, and atmospheric topped or reduced crude (ARC). In older refineries especially those that typically handle low sulfur crudes, the topped crude is sent to the VDU for separation into VGO and vacuum reduced crude (VRC) bottoms. However, modern refineries with high technology capable of processing crudes with high sulfur content typically employ an atmospheric residuum desulfurization (ARDS) unit for sulfur removal from the crude oil. Therefore, two design alternatives exist for ARC from ADU:

1. it is sent to the ARDS for sulfur removal to produce VRC, which is then sent to the VDU;
2. it is sent directly to the VDU to produce VGO and VRC, with the VGO subsequently hydrotreated for sulfur removal in a unit denoted as gas oil hydrotreater (GOHDT) (which stands for gas oil hydrotreater).

Thus, the associated logic proposition of design specification is stated as "select VDU only if ARDS is selected but without GOHDT selected." In other words, if ARDS exists in the optimal topology, then VDU must also exist but without the existence of GOHDT. The corresponding algebraic integer linear inequality and/or equality constraints in the optimization model are given by the following:

$$y_{\mathrm{ARDS}} \leq y_{\mathrm{VDU}} \tag{8.2}$$

$$y_{\mathrm{ARDS}} + y_{\mathrm{GOHDT}} = 1 \tag{8.3}$$

$$y_{\mathrm{GOHDT}} \leq y_{\mathrm{VDU}} \tag{8.4}$$

8.1.3.4.2 Processing Pool 2: Alternatives for Naphtha Exiting Hydrotreater/Hydrodesulfurizer

For the full-range naphtha leaving ADU that has been treated for sulfur removal via the hydrotreater (NHDT) or hydrodesulfurizer (HDS), the following alternatives are available:

- its subcomponent of the light straight-run naphtha (LSRN) stream from the top of the distillation column is sent to a gasoline blending pool (BLND);
- it is utilized as a feedstock for the catalytic reformer (CREF) and/or the isomerization unit (ISO);
- it is directly sold (SOLD).

Therefore, the corresponding constraint is given by:

$$y_{\mathrm{BLND}} + y_{\mathrm{CREF}} + y_{\mathrm{ISO}} + y_{\mathrm{SOLD}} \geq 1 \tag{8.5}$$

The same alternatives are applicable for the heavy straight-run naphtha (HSRN) after its subsequent treatment in another hydrotreater or hydrodesulfurizer.

8.1.3.4.3 Processing Pool 3: Alternatives for Vacuum Gas Oil Processing

The VGO stream is fed either to the fluidized catalytic cracker (FCC) or the hydrocracker (HCR) following hydrotreatment in GOHDT. Both FCC and HCR convert heavy gas oils into lighter products that are subsequently utilized as blendstocks for gasoline and diesel fuels. Hence, in general practice, both units do not coexist in a single site especially for relatively low-to-medium crude oil throughput unless the economies of scale as dictated by a high throughput justifies the routing of the hydrotreated VGO to be split into two streams, one for FCC and the other for HCR. Nevertheless, in principle, both units can coexist, with HCR usually favored over

FCC and is thus relatively more common, particularly in large-scale refineries that typically handle high crude oil throughput. Therefore, the constraint that allows at least one of these units to be selected is stipulated as follows:

$$y_{\text{FCC}} + y_{\text{HCR}} \geq 1 \tag{8.6}$$

8.1.3.4.4 Processing Pool 4: Alternatives for Vacuum Residue or Vacuum Reduced Crude Processing and Upgrading

Depending on the crude oil type and related process economics, VRC is further processed for production of transportation fuels (i.e. gasoline, kerosene, and diesel), typically via one of the following intermediary process units: visbreaker (VIS), solvent deasphalter (SDA), or mild hydrocracker (M-HCR). The corresponding constraint is to select none of these units (the provision for this option is the selection of an H-Oil unit, which is introduced later) or at most one unit among these three, as expressed by:

$$y_{\text{VIS}} + y_{\text{SDA}} + y_{\text{M-HCR}} \leq 1 \tag{8.7}$$

If none of the intermediate units or the delayed coker (COK) is selected, VRC is directly processed in the H-Oil unit (Kamiya 1991). Since both COK and H-Oil completely convert their feed material to extinction (100% conversion), they are not used in the presence of one another or other process units. Thus, a constraint that enables the selection of either one of these units is enforced as:

$$y_{\text{COK}} + y_{\text{H-OIL}} = 1 \tag{8.8}$$

8.1.3.5 Environmental Performance Assessment for Risk Evaluation of Flowsheets

To incorporate environmental considerations in the proposed modeling framework, we utilize the LCA approach proposed by Allen and Shonnard (2002) that uses certain performance assessment metrics for the environmental risk evaluation of process flowsheets. The methodology aims to rank the available design alternatives by performing their relative environmental risk assessment by integrating the following aspects into the design: (i) emissions estimation, (ii) environmental fate and transport calculations, and (iii) environmental impact data and indicators.

In this work, we represent the refinery air emissions with a set of relative environmental risk indices that measures the potential of global warming (GWP), stratospheric ozone depletion (ODP), acid rain deposition/acidification (ARP), and smog formation (SFP). To estimate the index for a particular impact category, we sum the contributions of each chemical released from a process weighted by their emission rate, yielding:

$$I_{\{\text{GWP,ODP,ARP,SFP}\}\in\text{DPRI}} = \sum_{i\in I} (\text{Dimensionless Potential Risk Index DPRI})_i \times m_i \tag{8.9}$$

in which the emission rate m_i is given by the multiplication of the emission factor and mass flowrate. The greenhouse chemicals or pollutants i considered in this work are CO_2, CO, SO_x, and NO_x.

To illustrate an example, the sum for product of the GWP and the mass emission rate of a pollutant over all pollutants considered, results in I_{GW} for the entire process, which in other words, is the sum of the emissions-weighted GWPs for each pollutant. It provides the equivalent process emissions of greenhouse chemicals in the form of the benchmark compound CO_2. The summation of the indices, as given by the following expression, is appended to the objective function for minimization:

$$\sum_{i \in I} \sum_{p \in P} (I_{\text{GWP},i} + I_{\text{ODP},i} + I_{\text{ARP},i} + I_{\text{SFP},i}) \tag{8.10}$$

8.1.3.6 Objective Function

Our goal is to determine the flowsheet of the optimal refinery network topology with the minimum annualized cost and environmental impacts. The objective function involves a combination of the following:

- minimizing the cost components that consist of the capital investment cost for equipment (CC_i), installation cost (IC_i), raw material cost (RMC_i), and operating cost (OC_i) associated with utility consumption (electricity, cooling water, and steam);
- maximizing revenues from the sales of refined products (S_i); and
- minimizing the environmental risk indices.

Thus, the objective function is expressed as:

$$\min z = \underbrace{\sum_{i \in I} (\text{CC}_i + \text{IC}_i + \text{RMC}_i + \text{OC}_i - \text{S}_i)}_{\text{economic-based costs}} + \underbrace{\sum_{i \in I} \sum_{p \in P} (I_{\text{GWP},i} + I_{\text{ODP},i} + I_{\text{ARP},i} + I_{\text{SFP},i})}_{\text{environmental risk indices}} \tag{8.11}$$

8.1.4 Numerical Example

We demonstrate the implementation of the MILP model formulation to determine the refinery topology on a numerical example using GAMS 22.3 under Integrated Development Environment (IDE) for Windows platform (Brooke et al. 1998). Data on the capital and operating costs of the process units are obtained from Maples (1993) and adjusted to the second quarter of year 2007 using the Marshall & Swift equipment cost index for petroleum products (*Chemical Engineering* 2007). As well, all monetary figures are adjusted to the rate of the US dollars for the year 2007. The model is solved using GAMS/CPLEX 10 (ILOG) on an Intel Pentium M processor 1.40 GHz, 496 MB of RAM laptop. The associated computational statistics are reported in Table 8.1.

The computational experiments conducted via the numerical example also serve to demonstrate that if an aggregated model is developed by keeping it as linear as possible and with a minimum use of binary variables, then the model can likely be solved to optimality by employing a simultaneous optimization strategy.

Table 8.1 Computational statistics on the problem size of the refinery topology MILP model.

Solver	Cplex 10
No. of continuous variables	573
No. of binary variables	16
No. of constraints	590
CPU time/resource usage	0.931 s
No. of iterations	64

8.1.5 Concluding Remarks

This work presents a superstructure optimization approach for synthesizing an oil refinery topology using an aggregated model to facilitate the preliminary screening stage of design alternatives. We are currently extending the model to incorporate more realistic representations of the refinery process flow, particularly by adopting more representative nonlinear models for the process units.

8.2 Industrial Case Study 2: Refinery Configuration with Heat Integration

This case study addresses the flowsheet optimization of the synthesis of a petroleum refinery to attain an optimal heat-integrated configuration or topology. A sequential two-step strategy is employed that first performs simultaneous flowsheet optimization and heat integration to obtain an optimal refinery topology with minimum utility cost. Subsequently, the fixed optimal topology with minimum utility loads is optimized to arrive at a configuration with the fewest heat exchanger units. A MILP is formulated based on a superstructure representation that considers many alternative feasible refinery topologies. The rest of the case study is organized as follows. Section 8.2.1 presents a formal description of the problem addressed in this work. The superstructure optimization-based modeling and computational strategy that is adopted for the synthesis problem are outlined in Section 8.2.2. Section 8.2.2 delineates the development of a superstructure representation as illustrated for the numerous processing alternatives of naphtha produced from an ADU. The complete formulation of the optimization model constraints and the objective function is then explained in Section 8.2.4. Section 8.2.5 discusses the computational results before some final remarks are provided.

8.2.1 Problem Statement

This example aims to determine an optimal heat-integrated topology or configuration of a petroleum refinery. Given are: a set of refinery process units with known yields and capacities; a set of process streams with known compositions with subsets of hot streams (require cooling) and cold streams (require heating), which have constant heat capacities and heat transfer coefficients; a set of refinery end-products with known market demands; the costs of the raw materials and prices of the saleable products; and the capital costs for purchasing and installation of the units. We wish to determine the optimal selection of the units (and their resulting interconnections); the stream flowrates; the required heating and cooling utilities; and the fewest potential stream matches in which each involves a hot stream exchanging heat with a cold stream that gives rise to a minimum number of countercurrent heat exchanger units (with their heat loads and operating temperatures).

8.2.2 Superstructure Representation

Figure 8.2 shows a state–task network (STN)-based superstructure representation (Al-Qahtani and Elkamel 2009, 2010; Khor and Elkamel 2010) that embeds many possible alternative refinery topologies as exemplified for the subsystem of naphtha produced from the ADU of a refinery. The first processing step in petroleum refining is crude oil distillation in which crude oil (CR) is distilled into oil fractions with respect to its boiling points. Naphtha constitutes the lighter fractions that are obtained from this process. Depending on the crude distillation column design and the refining economics, the ADU can produce a light straight-run naphtha (LSRN1) stream and a heavy straight-run naphtha (HSRN1) stream or an undifferentiated class of naphtha called wild naphtha (NAP1).

In the first instance, LSRN1 is mixed with purchased naphtha (PCHN2) and a LSRN2 stream from the hydrotreater (HDT1) in a mixer (MIX3). The output from MIX3, i.e. LSRN4, can either be used as a feedstock for the isomerization unit (ISO) or sold as a final product. Isomerization yields isomerate (ISO), one of the blending components for gasoline (GSLN).

On the other hand, HSRN1 is mixed with naphtha from the cracking of heavier fractions in a mixer (MIX1) before being sent to HDT1 to be desulfurized. HDT1 produces hydrogen sulfide gas (H2S1), liquefied petroleum gas (LPG1), desulfurized naphtha (LSRN2, HSRN3, and NAP4), and fuel gas (FG1). H2S1 is sent to a sulfur recovery unit (SRU) where sulfur (S) is extracted and finally sold while emitting a tail gas (TG) stream.

All LPG1, LPG2, and LPG3 streams are sent to a mixer MIX6 and subsequently to the LPG recovery unit (LPG), from which treated LPG (LPG5) is sold. Similar to the outputs from an ADU, the desulfurized naphtha from HDT1 can be classified as light (LSRN2) and heavy (HSRN3) or wild (NAP4). HSRN3 is sent to a mixer (MIX4), possibly with purchased naphtha (PCHN–3-1) and/or naphtha from the hydrocracker (HCR3).

The output of MIX4, i.e. a HSRN5 stream is the feedstock for the reformer (REF). FG1 goes to the FGH, supplying fuel gas (FG5) for the refinery. In the case that NAP4 is produced from HDT1, it is also mixed with purchased naphtha (PCHN3-2) and/or naphtha from the hydrocracker (HCR4) in MIX5, whose

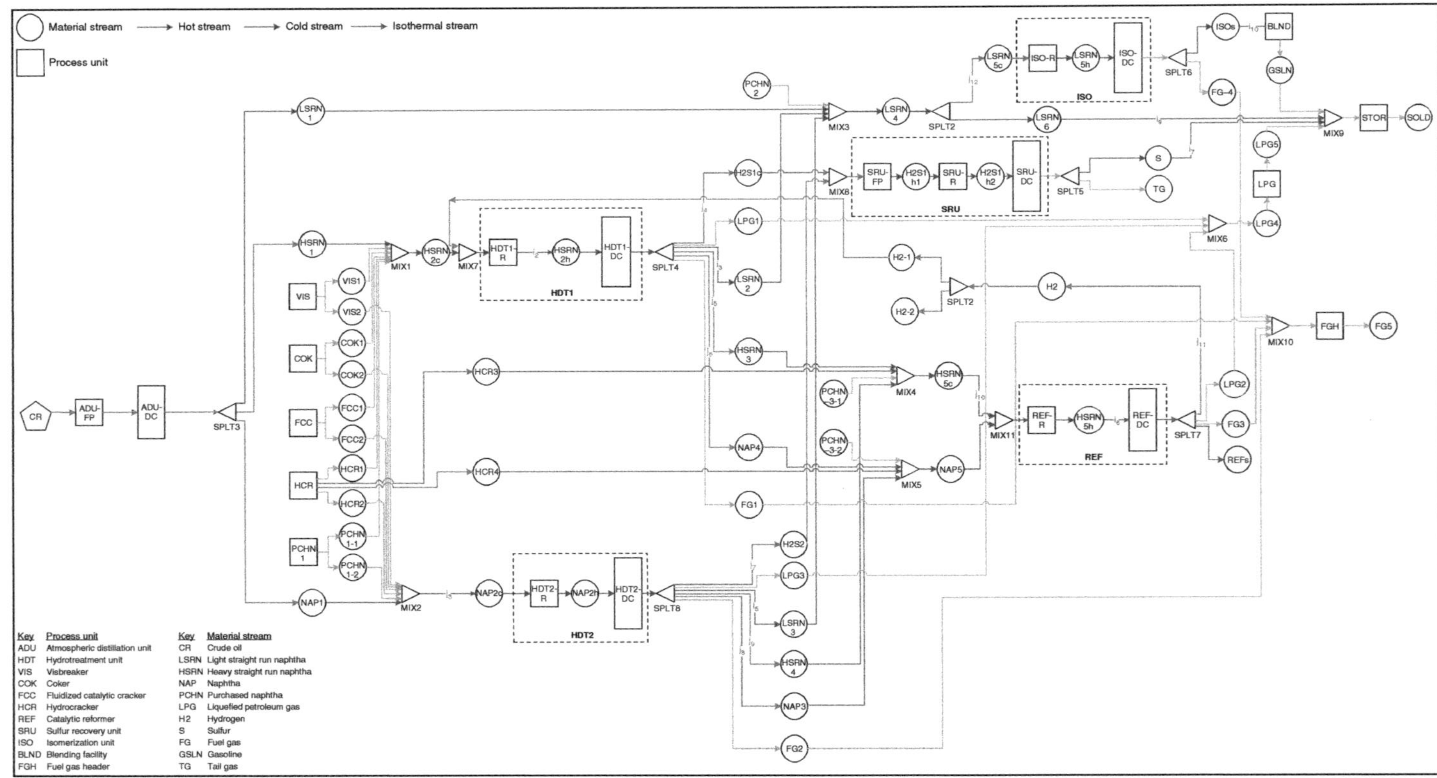

Figure 8.2 Superstructure representation for the naphtha produced from the atmospheric distillation unit. Source: Khor and Elkamel (2016). Reproduced with permission of John Wiley & Sons Inc.

output of NAP5 is sent to the reformer. The products from the reformer are hydrogen gas (H2), FG3, LPG (LPG2), and reformate (REFs). H2 is a feed to the HDTs, while reformate is used as a gasoline blending component. FG3 is sent to the FGH.

In the second instance involving NAP1 exiting ADU, the processing route is similar to the first instance in that NAP1 is mixed with naphtha from cracking processes in a mixer (MIX2) before being hydrotreated in HDT2. The products from HDT2 are H2S2, LPG3, the desulfurized naphtha of LSRN3, HSRN4, and NAP3, as well as FG2. Each product has the exact same route as the products from HDT1. Besides the ADU, naphtha is also produced from the cracking of distillation bottom products in a visbreaker (VIS), coker (COK), FCC, and hydrocracker (HCR). VIS has the lowest severity, while COK has the highest.

Due to the various differing temperature levels of the streams involved, there are significant opportunities for energy heat integration to take place. Such an instance involves supplying the heat available from the hot effluent stream of the reactor in the reforming section (HSRN5h) to its cold feed stream (HSRN5c) as well as to the cold stream entering the isomerizer (LSRN5c). The selection of the heat integration choice of stream matches is automated by evaluating the supply and target temperatures of the streams considered, thereby extracting the optimal configuration from the superstructure that embeds many (if not all) of the available design alternatives for processing naphtha from an ADU. The MILP-based transshipment formulation, which is presented in the next section, serves to optimally select the most preferred matches, which minimizes the total annualized refinery cost while catering for multiple utilities, allowing stream splitting, and obeying certain restricted matches due to practical plant considerations.

8.2.3 Modeling and Computational Strategy

We first perform flowsheet optimization simultaneously with heat integration considerations to minimize utility cost. A superstructure is postulated that considers multiple possible processing and heat integration alternatives, which give rise to many potential flowsheet structures or configurations. We employ 0–1 variables to represent the discrete decisions of selecting the available units and streams that constitute these numerous alternatives (Grossmann 2005). In doing so, we formulate a MILP that minimizes the total annualized processing and utility costs to obtain the optimal continuous variables on the stream flowrates with the associated process units selected and the utility loads. The underlying assumptions that allow for such a linear formulation are that the stream compositions are fixed as well as that of the temperature and pressure levels without dependence on the possibly nonlinear thermodynamics-related physical parameters such as the heat transfer coefficients to determine the heat fluxes involved. Subsequently, we optimize the number of heat exchanger units for the fixed optimal flowsheet structure with minimum utilities, in which each exchanger unit is associated with the heat recovery of a potential stream

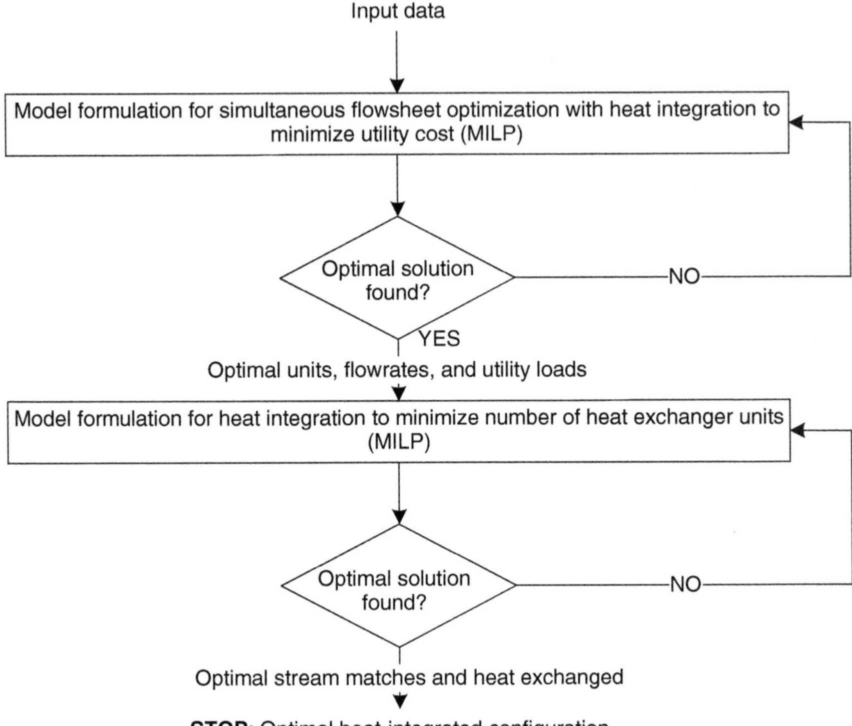

Figure 8.3 Flowchart of the modeling and computational approach. Source: Khor and Elkamel (2016). Reproduced with permission of John Wiley & Sons Inc.

match between a hot stream and a cold stream as represented by a 0–1 variable (Serna-González et al. 2007).

Figure 8.3 summarizes the foregoing procedure in the form of a flowchart. We adopt such a sequential optimization synthesis strategy that affords a large-scale refinery configuration problem to be addressed as a sequence of smaller problems. Nonetheless, the formulated MILP can account simultaneously for the structural (i.e. configuration) and parameter (i.e. the flowrates) optimization of both the process plant of the refinery and its associated heat recovery network. However, such an approach is at the expense of obtaining an improved solution by accounting for all the important tradeoffs simultaneously (Grossmann and Guillén-Gosálbez 2010).

8.2.4 Model Formulation

8.2.4.1 Flowsheet Optimization

To formulate a MILP model for the synthesis of a heat-integrated refinery for the developed superstructure, the following definitions are used:

(a) Sets and indices

I	Set of process units i
J	Set of process streams j
K	Set of temperature intervals k
$\mathrm{CR}(j)$	Set of crude oil feedstock j
$P(j)$	Set of end products j
$I(i, j)$	Set of inlet streams j into a process unit i
$O(i, j)$	Set of outlet streams j from a process unit i
$H(j, k)$	Set of hot process stream $h(j)$ that transfers heat at interval k or as heat that cascades from an interval higher than k
$C(j, k)$	Set of cold process stream $c(j)$ that receives heat at interval k
$M(m, k)$	Set of hot utility stream m that transfers heat at interval k or as heat that cascades from an interval higher than k
$N(n, k)$	Set of cold utility stream n that receives heat at interval k

(b) Parameters

$\mathrm{oc}_m^{\mathrm{HU}}$	Unit operating cost of hot utility m
$\mathrm{oc}_n^{\mathrm{CU}}$	Unit operating cost of cold utility n
$\mathrm{cc}_i^{\mathrm{f}}$	Fixed capital investment for process unit i
$\mathrm{cc}_i^{\mathrm{w}}$	Working capital for process unit i
CP_j	Heat capacity of stream j
$\mathrm{OC}_j^{\mathrm{f}}$	Fixed operating cost for stream j
$\mathrm{OC}_j^{\mathrm{v}}$	Variable operating cost for stream j
$\mathrm{OC}_j^{\mathrm{g}}$	General expenses
$T_{j,k}^{\mathrm{in}}$	Supply temperature for stream j at interval k
$T_{j,k}^{\mathrm{out}}$	Target temperature for stream j at interval k

(c) Continuous variables

F_j	Flowrate of stream j
Q_m^{HU}	Heat load of hot utility (HU) m
Q_n^{CU}	Heat load of cold utility (CU) n
$Q_{h,c,k}$	Heat exchange between hot stream h and cold stream c at interval k
$Q_{m,c,k}$	Heat exchange between hot utility m and cold stream c at interval k
$Q_{h,n,k}$	Heat exchange between hot stream h and cold utility m at interval k
$R_{h,k}$	Heat residual of hot stream h cascaded from interval k
$R_{m,k}$	Heat residual of hot utility m cascaded from interval k

(d) Integer 0–1 variables

$y_{h,c}$	Existence of match between hot stream h and cold stream c to exchange heat

8.2.4.1.1 Material Balances

Material balances with total stream flows around a process unit i:

$$\sum_{j\in I(i,j)} F_j = \sum_{j\in O(i,j)} F_j, \quad \forall i \in I \tag{8.12}$$

Material balances with fixed yields around a process unit i:

$$F_j = \sum_{j'\in I(i,j')} a_{i,j,j'} F_{j'}, \quad \forall i \in I, \forall j \in J \tag{8.13}$$

where $a_{i,j,j'}$ is a constant for process unit i that gives the distribution in stream $j \in O(i, j)$ coming from stream $j' \in O'(i, j')$.

8.2.4.1.2 Constraints on Production Requirements

A constraint is specified on the minimum ($F_j^{\mathrm{F,min}}$) and maximum ($F_j^{\mathrm{F,max}}$) levels of the raw material crude oil (CR) feed flow rate:

$$F_j^{\mathrm{F,min}} \le F_j^{\mathrm{F}} \le F_j^{\mathrm{F,max}}, \quad \forall j \in \mathrm{CR}(j) \tag{8.14}$$

8.2.4.1.3 Constraints on Production Levels

Minimum production levels must be satisfied for the end products to meet market demand d_j for product j:

$$F_j \ge d_j, \quad j \in P(j) \tag{8.15}$$

8.2.4.1.4 Constraints on Fuel Gas Requirement

A minimum FG requirement for internal plant consumption is enforced:

$$F_j^{\mathrm{min}} \le F_j, \quad j = \mathrm{FG} \tag{8.16}$$

8.2.4.1.5 Constraints on Import Requirement

Other than an atmospheric crude distillation unit (ADU), naphtha can also be sourced from the cracking of an ADU bottom products in a visbreaker (VIS), FCC, hydrocracker (HCR), and coker (COK), or purchased from the market (PCHN). Constraints are stipulated on the maximum import requirement h_j from an external naphtha source j as follows:

$$F_j \le h_j, \quad j = \{\mathrm{VIS, FCC, HCR, COK, PCHN}\} \tag{8.17}$$

8.2.4.1.6 Switching Constraints

The switching constraints are logical constraints to stipulate zero input flow for a nonselected unit or stream, which adopts a big-M formulation as follows:

$$F_j \le M_i y_i, \quad \forall (i,j) \in I(i,j) \tag{8.18}$$

8.2.4.1.7 Heuristic-Based Logical Constraints

We incorporate heuristic-based linear logical constraints in the formulation that enforce relevant design and structural specifications on refinery processes.

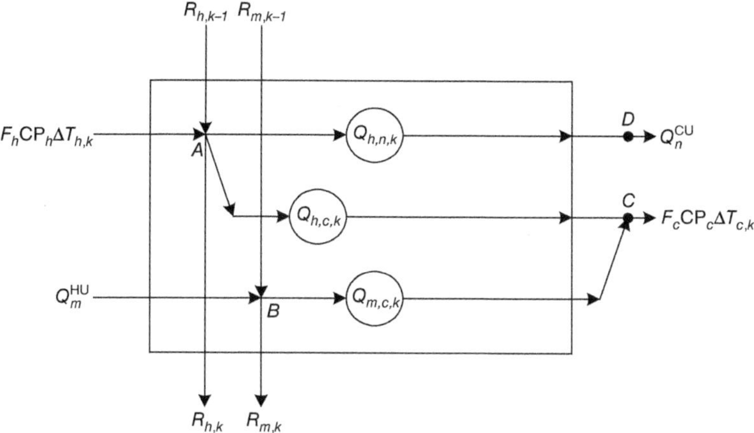

Figure 8.4 Superstructure representation on the heat flows in a temperature interval. Source: Khor and Elkamel (2016). Reproduced with permission of John Wiley & Sons Inc.

The constraints can be derived from propositional logic expressions that are developed based on past design experience, rules of thumb, and process engineering knowledge (Khor et al. 2011). The inclusion of these additional constraints may be able to enhance the solution convergence to optimality by removing undesirable solutions from the enumeration procedure (Khor et al. 2012).

8.2.4.2 Heat Integration Constraints

We employ the transshipment-based linear programming model of Papoulias and Grossmann (1983) to consider heat integration of the refinery systems to minimize the heating and cooling utility cost by maximizing the heat recovery of potential stream matches. The formulation requires partitioning the temperature range of the problem into intervals by assuming a fixed heat recovery minimum approach temperature. Figure 8.4 shows a superstructure representation on the heat flows in a temperature interval for which energy balances are formulated to relate the heat content, heat exchanges, and heat residuals as explained in the following.

Energy balance around a hot process stream $h(j)$:

$$R_{h,k} + \sum_{c\in C(j,k)} Q_{h,c,k} + \sum_{n\in N(n,k)} Q_{h,n,k} = R_{h,k-1} + F_h\text{CP}_h\left(T_{h,k}^{\text{in}} - T_{h,k}^{\text{out}}\right), \quad \forall (h,k) \in H(j,k) \tag{8.19}$$

Energy balance around a hot utility stream m:

$$R_{m,k} + \sum_{c\in C(j,k)} Q_{m,c,k} = R_{m,k-1} + Q_m^{\text{HU}}, \quad \forall (m,k) \in M(m,k) \tag{8.20}$$

Energy balance around a cold process stream $c(j)$:

$$\sum_{h\in H(j,k)} Q_{h,c,k} + \sum_{m\in M(m,k)} Q_{m,c,k} = F_c\text{CP}_c\left(T_{c,k}^{\text{out}} - T_{c,k}^{\text{in}}\right), \quad \forall (c,k) \in C(j,k) \tag{8.21}$$

Energy balance around a cold utility stream n:

$$\sum_{h\in H(j,k)} Q_{h,n,k} = Q_n^{\mathrm{CU}}, \quad \forall (n,k) \in N(n,k) \tag{8.22}$$

The following constraints stipulate zero heat residual from a previous higher temperature interval than the highest temperature interval at which a hot stream $h(j)$ or a hot utility stream m is present at:

$$R_{h,0} = R_{m,0} = 0. \tag{8.23}$$

Similarly, the following constraints stipulate zero heat residual from the last temperature interval:

$$R_{h,K} = R_{m,K} = 0. \tag{8.24}$$

It is noteworthy that multiple types of utilities and certain preferred or restricted stream matches can also be included in the formulation as desired.

8.2.4.3 Objective Function

The objective function of the model is to minimize the total annualized cost for processing and utilities as given by the following expression:

$$\min \sum_{i\in I} \mathrm{cc}_i^{\mathrm{f}} + \sum_{i\in I} \mathrm{cc}_i^{\mathrm{w}} + \sum_{j\in J} \mathrm{oc}_j^{\mathrm{f}} + \sum_{j\in J} \mathrm{oc}_j^{\mathrm{v}} + \sum_{j\in J} \mathrm{oc}_j^{\mathrm{g}} + \sum_{m\in M} \mathrm{oc}_m^{\mathrm{HU}} Q_m^{\mathrm{HU}} + \sum_{n\in N} \mathrm{oc}_n^{\mathrm{CU}} Q_n^{\mathrm{CU}}. \tag{8.25}$$

For the MILP formulation to minimize the number of heat exchanger units with fixed optimal topology and minimum utility load, the objective function is given by the sum of each of the possible stream matches between a hot stream $h(j)$ and a cold process stream $c(j)$:

$$\min \sum_{h\in H(j,k)} \sum_{c\in C(j,k)} y_{h,c} \tag{8.26}$$

where the 0–1 variable $y_{h,c}$ denotes the existence of such a match that can be associated with a single heat exchanger unit.

As before, the heat balances remain as the model constraints with the addition of the following logical constraints to ensure that no heat is exchanged if a match does not exist:

$$\sum_{k} Q_{h,c,k} \leq Q_{h,c,k}^{\mathrm{U}} y_{h,c}, \quad \forall h \in H(j,k), \forall c \in C(j,k) \tag{8.27}$$

where $Q_{h,c,k}^{\mathrm{U}}$ is an upper bound that can be stipulated as the smaller heat content for a pair of streams matched.

8.2.5 Computational Results

The proposed modeling and computational approach is applied to the synthesis problem for the processing of naphtha exiting an ADU as described through the

Table 8.2 Data on hot streams.

Hot stream	Supply temperature (°C)	Target temperature (°C)	Heat capacity flowrate (kJ/kg·°C)
LSRN1	88	40	236
HSRN2h	420	40	240
LSRN2	88	40	236
LSRN3	88	40	236
HSRN5h	400	40	240
REF	200	40	200
H2S1h	230	46	76
S	120	40	76
ISO	205	40	236

Source: Khor and Elkamel (2016). Reproduced with permission of John Wiley & Sons Inc.

Table 8.3 Data on cold streams.

Cold stream	Supply temperature (°C)	Target temperature (°C)	Heat capacity flowrate (kJ/kg·°C)
HSRN2c	179	370	240
NAP2	135	370	246
H2S1c	40	280	36
HSRN3	179	485	240
NAP4	135	485	246
H2S2s	40	280	36
HSRN4	179	485	240
NAP3	135	485	246
HSRN5c	179	505	240
H2	70	150	30
LSRN5	88	205	236

Source: Khor and Elkamel (2016). Reproduced with permission of John Wiley & Sons Inc.

superstructure in Section 8.2.2. The data for the problem is given in Tables 8.2–8.7 and the computational performance is summarized in Table 8.8.

The optimal topology with minimum total annualized processing and utility costs is shown in Figure 8.5 with the optimal stream flowrates indicated. The transshipment model involves 14 temperature intervals and determines the

Table 8.4 Process data.

Parameter	Value
API gravity of crude oil	44.6
Crude oil flow rate F_{CR}	100,000 kg/day $\leq F_{CR} \leq$ 150,000 barrel/day
Production of gasoline F_G	$F_G \geq$ 100,000 kg/day
Heat recovery minimum approach temperature ΔT_{min}	20 °C

Source: Khor and Elkamel (2016). Reproduced with permission of John Wiley & Sons Inc.

Table 8.5 Economics data (Maples 2000a).

Parameter	Value
Cost of crude oil	RM120/barrel = RM1006/m^3
Cost of naphtha	RM0.52/kg
Annual operating time	330 day/year
Nelson–Farrar Refinery Construction Index (NFRCI)	1241.7 (for January 1991); 2067.2 (for December 2008)

Source: Khor and Elkamel (2016). Reproduced with permission of John Wiley & Sons Inc.

Table 8.6 Data on utilities (Malaysian Industrial Development Authority (MIDA) 2013).

Type	Pressure (psig)	Temperature	Unit cost
Fuel oil	—	650	RM0.1018/MJ
High-pressure steam	460	320	RM0.0050/kg
Medium pressure steam	230	200	RM0.0030/kg
Low-pressure steam	55	145	RM0.0015/kg
Cooling water	0	25	RM0.8400/m^3
Electricity	—	—	RM0.20/kWh

Source: Khor and Elkamel (2016). Reproduced with permission of John Wiley & Sons Inc.

presence of a single pinch point in the problem (corresponding to the temperatures 120 °C on the hot side and 110 °C on the cold side in an associated heat cascade diagram). This heat-integrated topology with minimum utility registers a total annualized cost of MYR2169 per year, which entails a 13% and 27% reduction as compared to the non-heat-integrated solutions of the same problem computed by the two techniques previously reported by the same author (Khor et al. 2011). The subsequent MILP model to minimize the number of heat exchangers yields 12 process-to-process exchangers, 6 hot utility exchangers, and 8 cold utility

Table 8.7 Base cost and utilities consumption of the major process units (Maples 2000b).

Process unit	Base cost		Utility consumption			
	January 1991 (RM million)	December 2008 (RM million)	Electricity (MWh/kg)	Fuel (kJ/kg)	Steam (kg/kg)	Cooling water (m^3/kg)
ADU	137	228	0.0039	0.0826	0.0888	0.0000
VIS	86	144	0.0039	0.0660	0.1776	0.0000
COK	166	276	0.0282	0.0991	0.1421	0.0000
FCC	310	515	0.0078	0.0660	0.0710	0.0119
HCR	342	569	0.1402	0.2766	0.0000	0.0000
HDT	58	96	0.0157	0.0248	0.0533	0.0000
REF	162	270	0.0078	0.2477	0.1421	0.0030
ISO	25	42	0.0078	0.0083	0.1279	0.0000
SRU	18 (per ton)	30 (per ton)	0.3132	0.0000	2.6636	0.1482

Source: Khor and Elkamel (2016). Reproduced with permission of John Wiley & Sons Inc.

Table 8.8 Model size and computational statistics.

Platform	GAMS 24.2.2
Solver	CPLEX
No. of single continuous variables	2513
No. of binary variables	152
No of constraints	304
CPU time	<1 s
Constraint satisfaction tolerance	10^{-5}
Relative optimality tolerance	10^{-1}
Absolute optimality tolerance	0

Source: Khor and Elkamel (2016). Reproduced with permission of John Wiley & Sons Inc.

exchangers as shown in Figure 8.6. (Note that there is a possibility to obtain a different network configuration with the same utility load and number of units).

8.2.6 Concluding Remarks

This work addresses the process synthesis problem to optimally determine a heat-integrated refinery topology. We employ a two-step procedure of flowsheet optimization and heat integration that involves the sequential minimization of the utility cost and the number of heat exchanger units. The approach is illustrated for the processing alternatives of a naphtha mixture exiting an ADU with computational results that show potential reduction in the total annualized cost.

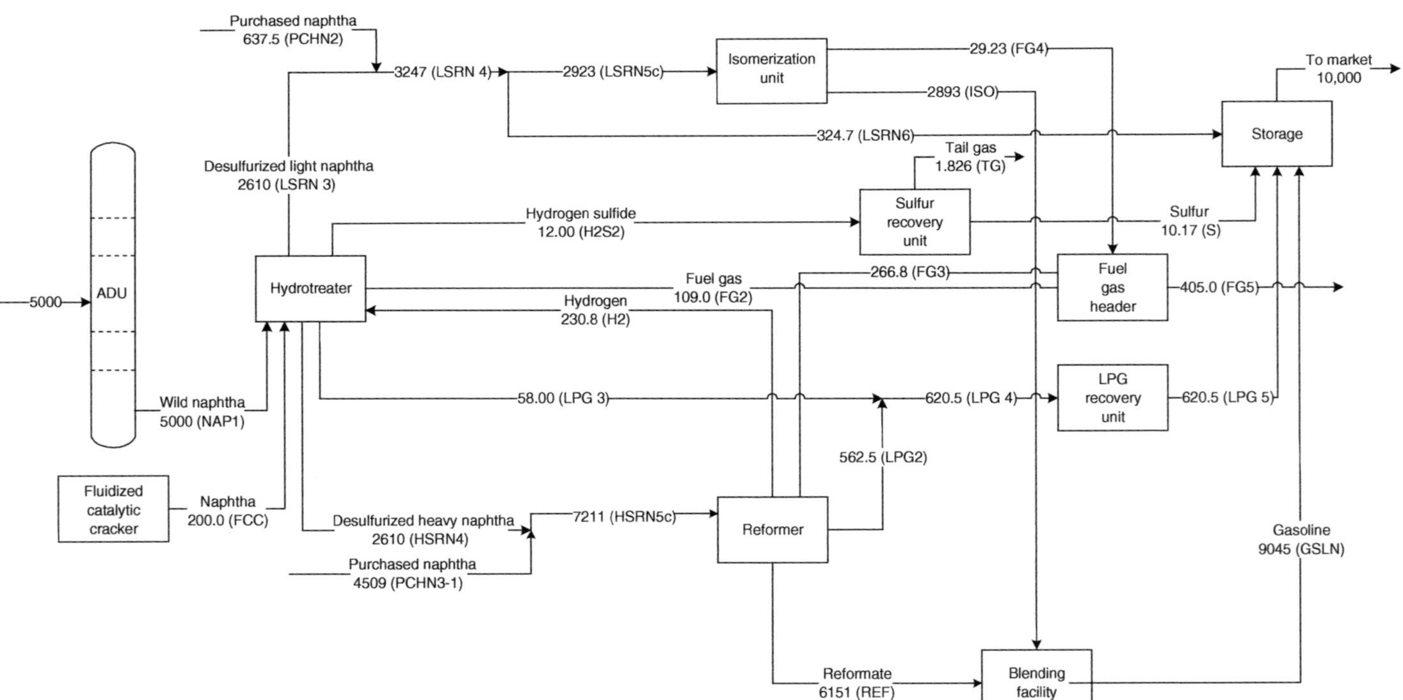

Figure 8.5 Optimal flowsheet with minimum utility costs for naphtha produced from atmospheric distillation unit. Source: Khor and Elkamel (2016). Reproduced with permission of John Wiley & Sons Inc.

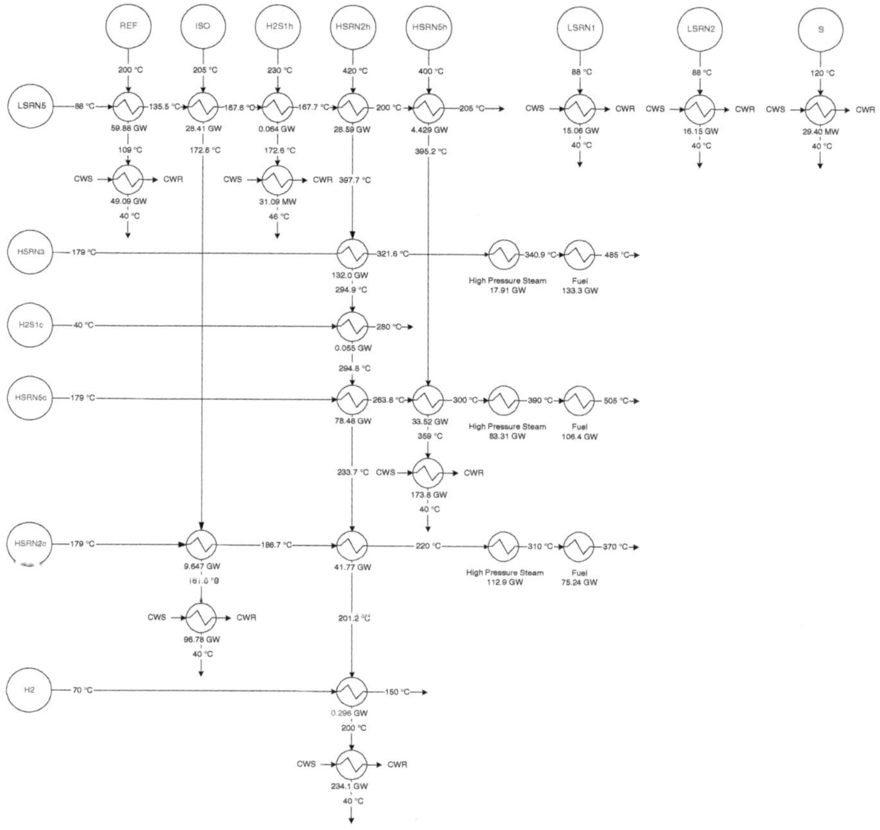

Figure 8.6 Optimal heat exchanger network with fewest units for naphtha produced from atmospheric distillation unit. Source: Khor and Elkamel (2016). Reproduced with permission of John Wiley & Sons Inc.

It is noteworthy that while cost reduction can be achieved by implementing our proposed approach of optimizing a plant topology to attain an optimal configuration from an energy standpoint, there are possible drawbacks particularly in terms of operating such a highly heat-integrated site. The operation challenge faced may actually reverse the cost advantage offered if, for instance, a heat-integrated stream could not be used due to an unexpected cause, which thus affects other related streams leading to an uneconomic situation of plant shutdown.

8.3 Chapter Summary

This chapter presents several applications of the foregoing optimization modeling framework introduced in earlier chapters. Four case studies on configuration design are illustrated on problems of industrial size and significance for the petroleum refining industry. Each of the applied examples emphasizes differing aspects of importance to this process synthesis problem type.

References

Al-Qahtani, K. and Elkamel, A. (2009). Multisite refinery and petrochemical network design: optimal integration and coordination. *Industrial & Engineering Chemistry Research* 48 (2): 814–826.

Al-Qahtani, K. and Elkamel, A. (2010). Robust planning of multisite refinery networks: optimization under uncertainty. *Computers and Chemical Engineering* 34: 985–995.

Douglas, J.M. (1985). A hierarchical decision procedure for process synthesis. *AIChE Journal* 31 (3): 353–362.

Gary, J.H., Handwerk, G.E., and Kaiser, M.J. (2007). *Petroleum Refining: Technology and Economics*, 5e. New York: Marcel Dekker.

Grossmann, I.E. (2005). Advances in logic-based optimization approaches to process integration and supply chain management. *Chemical Engineering Trends and Developments* 299–322.

Grossmann, I.E. and Guillén-Gosálbez, G. (2010). Scope for the application of mathematical programming techniques in the synthesis and planning of sustainable processes. *Computers & Chemical Engineering* 34 (9): 1365–1376.

Kamiya, Y. (1991). *Heavy Oil Processing Handbook*. Japan: Research Association for Residual Oil Processing (RAROP).

Khor, C.S. and Elkamel, A. (2010). Superstructure optimization for oil refinery design. *Petroleum Science and Technology* 28 (14): 1457–1465.

Khor, C.S. and Elkamel, A. (2016). Optimal synthesis of a heat-integrated petroleum refinery configuration. *The Canadian Journal of Chemical Engineering* 94 (10): 1939–1946.

Khor, C.S., Yeoh, X.Q., and Shah, N. (2011). Optimal design of petroleum refinery topology using a discrete optimization approach with logical constraints. *Journal of Applied Sciences* 11 (21): 3571–3578.

Khor, C.S., Chachuat, B., and Shah, N. (2012). A superstructure optimization approach for water network synthesis with membrane separation-based regenerators. *Computers & Chemical Engineering* 42: 48–63.

Linnhoff, B. (1993). Pinch analysis - A state-of-the-art overview. *Transactions of the Institution of Chemical Engineers* 71: 503–522.

Malaysian Industrial Development Authority (MIDA) (2013). *Malaysia Investment Performance 2012*. Kuala Lumpur: Government of Malaysia.

Maples, R.E. (2000a). *Petroleum Refinery Process Economics*, 2e, 264. Oklahoma: Pennwell.

Maples, R.E. (2000b). *Petroleum Refinery Process Economics*, 2e, 388. Oklahoma: Pennwell.

Papoulias, S.A. and Grossmann, I.E. (1983). A structural optimization approach in process synthesis-III. Total processing systems. *Computers and Chemical Engineering* 7 (6): 723–734.

Raman, R. and Grossmann, I.E. (1991). Relation between MILP modelling and logical inference for chemical process synthesis. *Computers and Chemical Engineering* 15 (2): 73–84.

Raman, R. and Grossmann, I.E. (1992). Integration of logic and heuristic knowledge in MINLP optimization for process synthesis. *Computers and Chemical Engineering* 16 (3): 155–171.

Raman, R. and Grossmann, I.E. (1993). Relation between MILP modeling and logical inference for chemical process synthesis. *Computers and Chemical Engineering* 15: 73–84.

Raman, R. and Grossmann, I.E. (1994). Modelling and computational techniques for logic based integer programming. *Computers and Chemical Engineering* 18 (7): 563–578.

Serna-González, M., Jiménez-Gutiérrez, A., and Ponce-Ortega, J.M. (2007). Targets for heat exchanger network synthesis with different heat transfer coefficients and non-uniform exchanger specifications. *Chemical Engineering Research and Design* 85 (10): 1447–1457.

9

Industrial Case Studies with Engineering-Centric Techno-Commercial Considerations

Two industrial case studies on refinery design configuration focusing on engineering-centric techno-commercial considerations are discussed in this chapter.

9.1 Industrial Case Study 1: Refinery Configuration for High-Octane Fuel Production

The global drive for environmental sustainability necessitates continuous adjustment, optimization, and improvement in petroleum refining processes to generate energy and products that include automotive fuels such as gasoline. At the same time, refiners need to maximize their asset utilization to maintain competitiveness in a business setting. This case study presents a process advisory and monitoring application to optimize a catalytic naphtha reforming operation to produce high-octane gasoline feedstock from a reformate product stream. A mathematical model is developed for the catalytic reforming process to produce high rates of hydrocarbons with excellent anti-knock ratings. The proposed methodology involves formulating a nonlinear programming optimization model to perform data reconciliation. The model objective serves to minimize the measurement errors in terms of deviations between the measurement values and the model reconciled values with suitable weighting strategies to reflect the accuracy and reliability of the measurements. The overall optimization procedure is carried out subjected to various operating constraints that meet important real-world conditions and equipment specifications, including compliance with environmental regulations to ensure sustainable processing of the required products, which include hydrogen gas and aromatics. We present a numerical example to illustrate an implementation of the resulting model in an online environment to improve process operation at an actual refinery in Canada. The computational results show enhanced product quality of a reformate stream with high octane number and increased yields.

Model-Based Optimization for Petroleum Refinery Configuration Design, First Edition. Cheng Seong Khor.

9.1.1 Catalytic Reforming Process

Catalytic reforming is an important process in the petroleum refining industry. It was originally developed to produce components of automotive fuels, specifically gasoline, which meet engine requirements for high anti-knock quality. The objective of the process is to convert petroleum naphtha fractions to high rates of aromatic hydrocarbons as selectively as possible since the latter have excellent antiknock ratings. Naphtha fractions are liquid hydrocarbon mixtures with 6–12 carbon atoms and boiling points in the range of 320–470 K (Moser and Sadler 2002; Moser and Bogdan 2008). Reforming also serves two other main purposes in a refinery: it is a main hydrogen producer for use within a refinery or outside it; it also provides feedstock (mainly consists of benzene, toluene, and xylene) for the subsequent downstream petrochemical production processes (Sinfelt 1981). A survey of recent progress on the reforming process focusing on the reactor modeling is available in reference (Rahimpour et al. 2013).

A commercial reformer unit consists of a reactor section, a recycle gas compression section, and a fractionation section. The reactor section consists of a feed system, a few heaters or furnaces, a series of reactors, and a flash separator. A portion of the flashed hydrogen gas is recycled and mixed with the feed, then increased to the reaction temperature by a heat exchanger combining the feed and reactor effluent followed by the first heater. We send the flashed liquid to the fractionation section, which comprises a distillation column that acts as a product stabilizer. The distillate stream strips light gases from the flashed liquid that produce liquefied petroleum gas (LPG) and off-gas as fuel. The main product is the bottom stream called reformate that is a feedstock for gasoline blending. We use a few (typically three to five) heater-and-reactor pairs to maintain the reaction temperature within a range of 400–500 °C (700–800 K) and at pressures of 10–35 atm using catalysts mainly to accelerate the reaction (Sinfelt 1981). Figure 9.1 displays a general process flow of a commercial reformer unit.

9.1.2 Data Reconciliation Method

Data reconciliation involves estimating process variables by comparing their values from process measurements and process models. The models typically comprise material and energy balances or conservation relations. Data reconciliation involves adjusting or correcting errors in measurement values to be consistent with the mass balances. The procedure is useful in the process and its affiliated industries to conduct monitoring and modeling for control, simulation, and optimization, as well as to perform instrument maintenance and equipment analysis (Sinfelt 1981; Amand et al. 2001; Heyen and Kalitventzeff 2006; Moser and Bogdan 2008; Rahimpour et al. 2013).

Progress has been recorded in the literature on data reconciliation techniques that include calculating measurement errors to estimate the model variable values using linear process models (Wu et al. 2016), classifying the estimated variables (Stanley and Mah 1981), and categorizing the measurement values as redundant or nonredundant (Crowe 1989). In this case study, we adopt the data reconciliation procedure

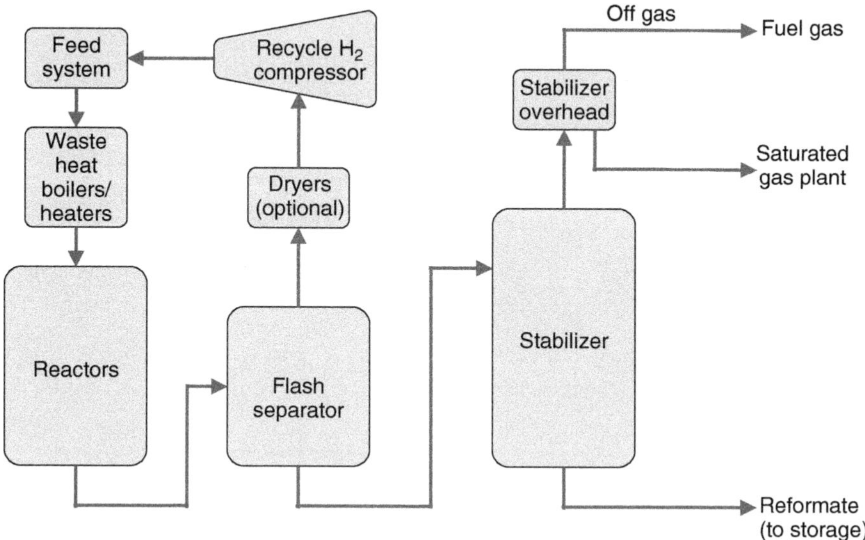

Figure 9.1 General process flow of a commercial reformer unit. Source: Khor (2019). Reproduced with permission of IntechOpen.

that makes use of information from redundant measurement and conservation laws to correct measurement values by converting them into model values based on reliable, accurate, and thermodynamically valid process knowledge incorporated in a nonlinear steady-state model.

9.1.3 Problem Statement

We have the following data for a catalytic reforming process: a set of process units; a set of measurements i with vector of known data on their raw scan values y, model values x, standard deviation σ_i, and weights w_i; and a set of tuning parameters p with known nominal values n_p, model values t_p, and scaling factors s_p. The model constraints for the process are mass balances, energy balances, and equilibrium relationships.

In this case study, we want to optimize the operation of a catalytic reforming process by minimizing the deviations of values for selected measurements and tuning parameters from their model-computed values (i.e. differences computed between the x and y vectors and the n_p and t_p vectors, respectively) subject to process constraints to obtain reconciled process values (as given by the model values). To achieve this aim, we can adjust a set of reconciled variables such as skew factors on the true boiling point of crude oil fractions, product flow rate bias, Murphree efficiencies of the distillation columns, and reaction kinetic parameters.

9.1.4 Model Formulation

This section describes the approach used to develop a reforming process monitoring model using the commercial advanced chemical process optimizer platform called

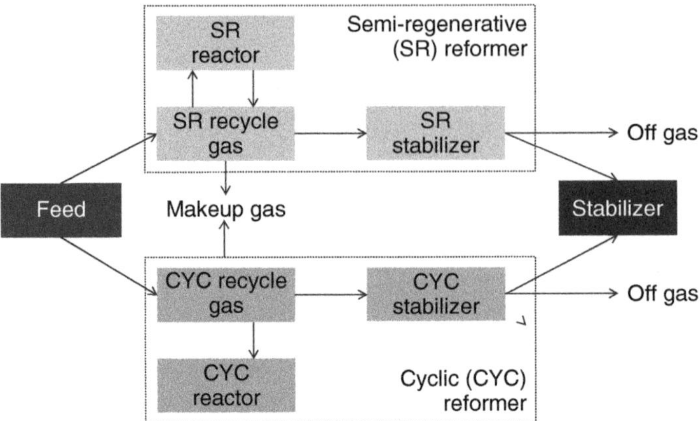

Figure 9.2 Model scope comprising the semi-regenerative and cyclic reforming units. Source: Khor (2019). Reproduced with permission of IntechOpen.

SimSci ROMeo (Schneider Electric 2017). The model developed involves two technologies: cyclic and semi-regenerative reforming with the overall scope covered by the model shown in Figure 9.2. Figure 9.3 presents a detailed schematic of the model for the cyclic reforming operation.

The optimality of a catalytic reforming process operation is measured based on the yields of the main product of reformate (a gasoline blending feedstock) and the side products of petrochemicals comprising benzene, toluene, and xylene (BTX). We use material-balanced data of the model-computed values and not raw scan values of the measurements to evaluate such performance measures. Hence, we perform data reconciliation on the measurement values, which are obtained from field instruments measuring the actual process operation.

9.1.4.1 Data Reconciliation Model

We formulate the data reconciliation procedure as a least squares minimization version of an optimization problem (Gill et al. 1981). The objective function as given in Eq. (9.1) minimizes the sum of squares of the weighted differences between model values and raw scan values for selected measurements and tuning parameters.

$$\text{minimize} \sum_i w_i \left(\frac{\delta_i}{\sigma_i} \right)^2 + w_p \left(\frac{t_p - n_p}{s_p} \right)^2 \tag{9.1}$$

where w_i is the weight matrix for each measurement i (either 1 or 0 otherwise for those not considered) and δ_i is the offset as given by the difference between the model value and the raw scan value.

The differences or deviations are also called errors, offsets, or biases of the instruments considered. The offset value is weighted or multiplied with a weight factor (i.e. its value is $w_i = 1$) if we decide to reconcile the process value of the associated measurement or tuning parameter. Such a decision is made based on a measurement's reliability, i.e. if its measured value is accurate to be compared against a model-reconciled value; otherwise the instrument needs to be calibrated. The model

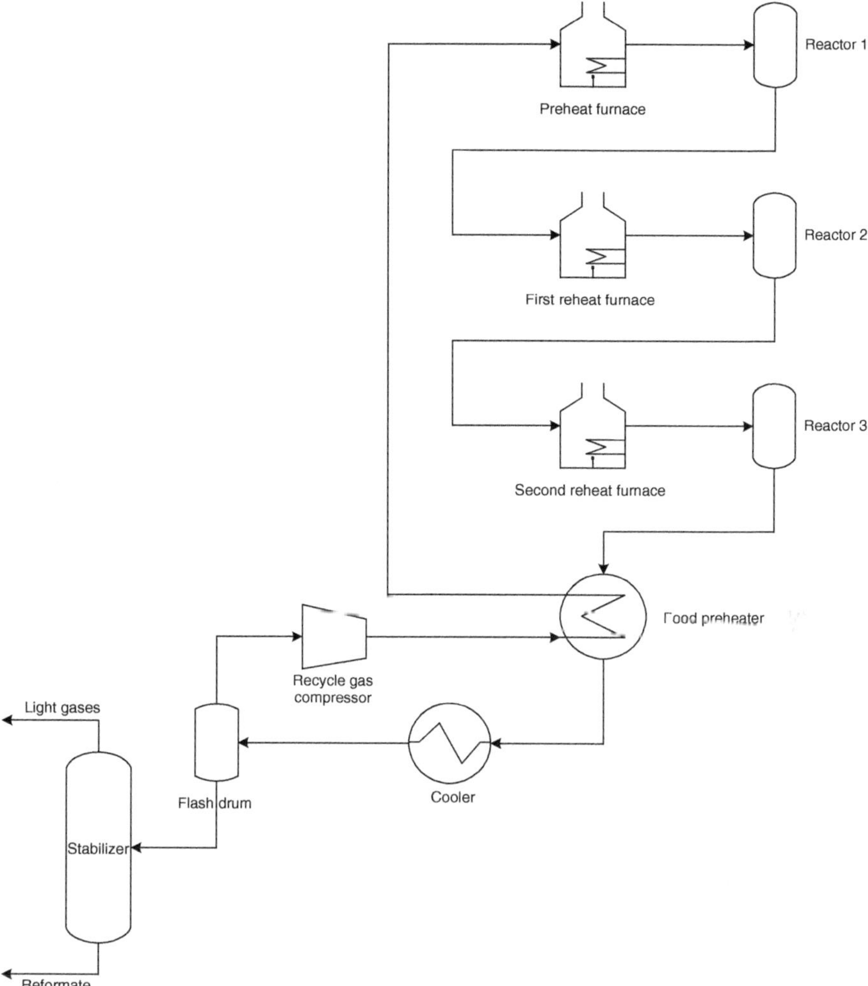

Figure 9.3 Schematic representation of the cyclic reforming operation model. Source: Khor (2019). Reproduced with permission of IntechOpen.

constraints mainly consist of the total mass and component balances, energy balances, and equilibrium relationships, as well as appropriate bounds on the variables (Wang and Romagnoli 2005; Lid and Skogestad 2008).

9.1.4.2 Feed Characterization

We perform tuning on the reformer feed composition to match the process data as closely as possible by adjusting the factors representing the component mole fractions (Britt and Luecke 1973). It is not necessary to adjust for all the components in the reconciliation procedure, especially if regular laboratory data on feed composition are not available in terms of its PONA (paraffins, olefins, naphthenes, and aromatics) content analysis. We can reduce the number of components to be tuned for the feed slate by equating the mole fraction factors for components with the same number of carbon atoms on the basis that these components have similar molecular weights and densities. This strategy can result in reducing about one-third of the parameters involved in the tuning procedure.

In the absence of composition data, we consider tightening the reformate yield specification by minimizing its offset and reducing its standard deviation value. The measured reformate yield value is given by the ratio between reformate mass flow rate and that of the feed; therefore, no weight is specified on the reformate yield because the actual measured values are the product and feed flows.

In a Component Adjuster model of the SimSci ROMeo software, three choices of methods are available for handling each component: Flow, Component, and Dependent. Typically, only one component is chosen as a Dependent component; all others should be chosen as a Flow or Component. There is no definite guideline on specifying the use of a Flow or Component method for a component. It is best to use the most abundant component as a Dependent component.

Depending on the situation, we can impose bounds on the adjustment factor to get a feasible solution. If the lower bound of the adjustment factor for a component is active, setting its lower bound to a value of −0.5 results in reducing the component mole fraction by 50% (e.g. from 0.05 to 0.025). Correspondingly, a lower bound of −1 reduces the component mole fraction to zero, i.e. the component is removed from the resultant outlet stream of the Component Adjuster. This effect is undesirable; we do not want a component to be removed completely. Therefore, a suggestion is to use a lower bound other than −1 (e.g. −0.8 or −0.75). Similarly, for a component with an active upper bound, an upper bound of 0.5 increases its mole fraction by 50% (e.g. from 0.05 to 0.075). Therefore, we avoid upper bounding the adjustment factor on the mole fraction of a component as 1 because it results in a doubling effect.

It is extensive and not necessary to tune the compositions of all components in the lumped feed slate. In particular, there is no need to adjust those of the light components with five carbon atoms (denoted as C5s) and lower (i.e. C3s to C5s) because their compositions are very low (or almost zero). A typical bound for the tuning parameters of a feed component is [−0.75, 0.75]; a practical bound depends on the actual data.

9.1.4.3 Reactor Representation

The application uses a proprietary nonlinear surrogate model to represent a reactor bed that incorporates its catalyst- and kinetics-related information. We use a flash drum operating under adiabatic condition to model the pressure of the reactor bed inlet stream and a similar configuration to model the outlet stream, as shown in Figure 9.4. For each reactor, we include a set of equations to equate each of the kinetic parameters (e.g. catalyst activity) to a common variable, thereby facilitating the reaction kinetic tuning to match the plant values. Since each reactor is identical, we first develop an individual reactor with its associated auxiliary units (such as the two flash drums at its inlet and outlet) and then duplicate its representation. This approach can be done by using a block diagram (or other similar feature) that is

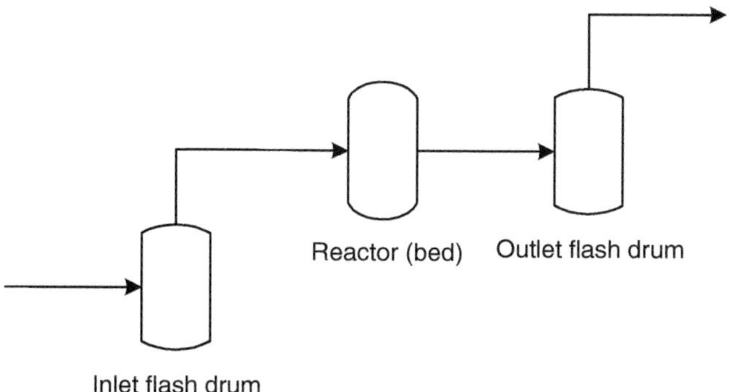

Figure 9.4 Schematic representation of a reactor. Source: Khor (2019). Reproduced with permission of IntechOpen.

available, which can help improve the layout of the graphical user interface besides facilitating a failed solution or modeling problem.

9.1.4.4 Reactor Pressure Balance

There are a number of ways to represent a reactor pressure balance for data reconciliation purpose depending on the reliability of measurements available: (i) distribute the total pressure drop across the reactor bed equally to the pressure drop of the inlet flash drum and that of the outlet flash drum; (ii) honor the inlet and outlet pressure measurements if they are reliable; and (iii) relate the pressure drops for the other reactors to that of a reactor with a known reliable pressure measurement.

9.1.4.5 Reaction Kinetic Tuning

A general approach in reforming applications is to maintain most of the built-in reaction kinetic tuning parameters equal for each of the reactors by using some form of mathematical relation facility (such as a customization unit feature available in ROMeo). We use the first mathematical relation with a local variable on each kinetic parameter that has an associated tuning parameter to adjust it. Then for each reactor, we use the second relation to equate a kinetic parameter to its corresponding variable in the first relation. Such a setup gives the flexibility of turning a reactor on or off when addressing convergence issues while keeping the kinetic tuning parameters equal for all reactors that are turned on. We put suitable bounds on the tuning parameters to keep the solved parameter values within a reasonable range. The tuning parameters are divided into two sets: for catalyst and for catalyst coke. The catalyst tuning parameters are base-2 logarithm ($\log_2$) multipliers that are bounded.

9.1.4.6 Reactor Switch in Cyclic Reformer

The application has a logic to represent the reactor switch for a cyclic reformer. The switch operation involves a swing reactor besides the on-oil reactors to replace the reactor in regeneration. Thus, the catalyst can be regenerated without shutting the unit down. Note that such logic is not required for a semi-regenerative reformer

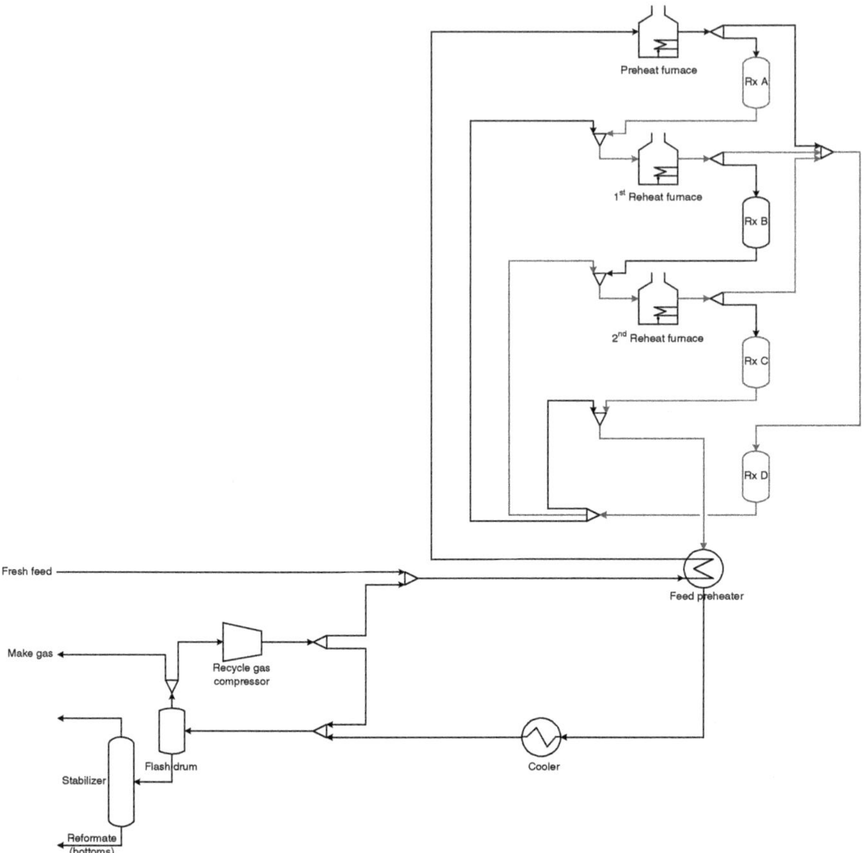

Figure 9.5 Reactor switching in cyclic reformer. Source: Khor (2019). Reproduced with permission of IntechOpen.

in which all the reactors are taken off-stream or out of service for in situ regeneration of the entire catalyst inventory.

The mentioned logic to model the reactor switching involves the following steps with an explanation aided by Figure 9.5:

- turns down the following items for the off-oil reactor and turns them on for the on-oil reactor: inlet pressure measurement, reactor custom model unit, inlet and outlet flash drums, customization units on reaction kinetic tuning, customization unit on reactor pressure balance, and delta temperature measurement model and its associated customization unit;
- sets to zero for the initial and final values of the off-oil reactor bed pressures (and equates these values to those of the inlet flash drum pressures for the on-oil reactors);
- sets to zero for the split fraction of the off-oil reactor inlet flow rate; sets to 1 for the split fraction of the swing reactor bypass flow rate and vice versa for the on-oil reactors; and thereafter generates estimates for these splitters.

If the swing reactor is off oil, the logic performs the following steps:

- turns on the inlet mixer and outlet splitter of the swing reactor;
- initializes the mixer pressure drop to that of the incoming bypass stream thereafter, generating value estimates;
- sets the split fractions for the swing reactor outlet splitter flow rates thereafter, generating value estimates;
- initializes the mixer pressure drop for each on-oil reactor to the incoming stream thereafter, generating value estimates;
- unweight the inlet pressure and temperature difference measurements.

The default setting of the logic is to put weights on all the reactor inlet pressure and temperature difference measurement models to account for them in the objective function computation. But when the reactor switch happens, the swing reactor inlet pressure measurement may have a large offset that results in a huge objective value; hence, it may merit resolving the model by removing the weight on this tag.

9.1.4.7 Measurement Models

9.1.4.7.1 Absolute Error Tolerance

The standard deviation parameter is specified to set the absolute error tolerance for a measurement model. We refrain from manipulating this tolerance to obtain a desired solution or to represent the importance of a measurement; instead, the tolerance represents a measurement's reliability and accuracy.

We use Microsoft Excel to view and analyze the results to help with obtaining a good representation of the measurements; an example of such a results viewer is shown in Figure 9.6. A measurement may still be of good quality but a particular instance of its value may be bad, for example, because data have stopped flowing to the data historian. Specifying the tolerance of a measurement depends on its

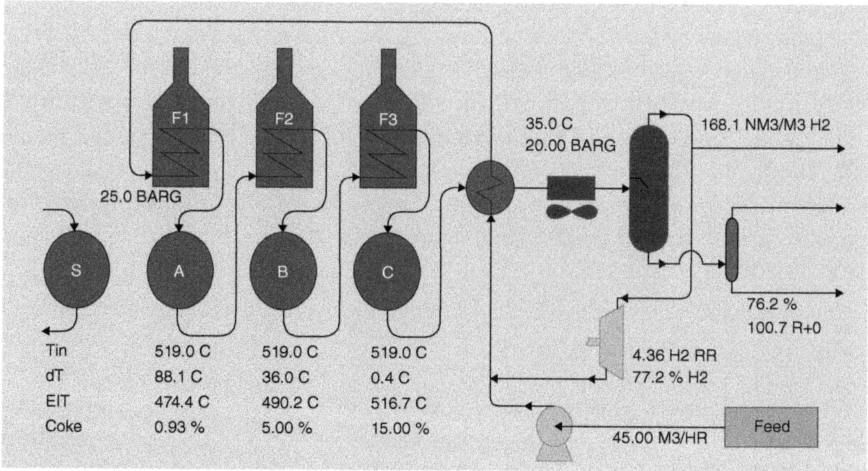

Figure 9.6 An example of results viewer of the reformer model. Source: Khor (2019). Reproduced with permission of IntechOpen.

precision. The weighting of a measurement can handle a bad value measurement instance by way of excluding it from being considered in the objective function computation.

It is recommended to set approximately 1% of the process variable (PV) value for most measurements and 5% of the PV for flow rates. Note that a calculated tag is generally not reconciled, i.e. unweighted if the input measurement for which the calculations are based on is available.

9.1.4.7.2 Measurement Screening Logic to Handle Data Quality

Specifying the logic for measurement model screening requires review when handling negative flow rate instances to determine if they are caused by zero flows or bad transmitters. The following gives a list of rules of thumb that can be applied:

- The typical response to a bad quality value is to use the last good raw scan value, which applies particularly for a measurement model whose value is a dependent variable. For temperature indicators (thermocouples), we can select to set their objective function contribution to zero.
- If we do not want a measurement value to become negative when its quality turns bad, we can fix its value to zero or a small value by setting its minimum and selecting an out-of-range action. This rule typically applies to a flow rate meter which we do not need to specify a maximum value.
- Set zero objective function contribution for the offset of a dependent-valued measurement.
- To respond to bad quality data for an independent-valued measurement, we can use a suitable fallback value as the scan value or stop running the online model to find out the cause, especially if this happens to a feed flow rate meter.
- To fix controller outputs at certain minimum and maximum values.

9.1.4.7.3 Two Measurements on One Variable

There may be two measurements available on the same process variable. It is acceptable to use two measurement models for a dependent variable, but for an independent variable, doing so results in one measurement model being used as a controller and another as an indicator, thus we may face problems if its scan value becomes bad. If both measurement models are necessary, we can specify a new variable equal to the particular PV and point the indicator measurement to it, besides specifying a screening criterion to set the objective value contribution to zero when the measurement data quality turns bad.

9.1.5 Results and Discussion

9.1.5.1 Key Process Variables

The key process variables to match are reformate yield, reformer reactor total endotherm, research octane number (RON) of reformate product, and hydrogen recycle gas purity. The available tuning variables are feed composition and reaction kinetic parameters such as overall catalyst activity, acid site activity, and aromatics

Table 9.1 Representative key process variable values to match error tolerance for reformate yields.

Variable	Error	Value (offset in bracket)	
		Cyclic	Semi-regenerative
Reformate yield	±0.1%	81.19 (−0.13)%	82.68 (−0.26)%
Reactor total endotherm	±10°F	198.17 (−0.01)°F	152.07 (−5.20)°F
Research octane number (RON)	±0.1	87.39 (−0.35)	87.53 (−0.63)
Hydrogen recycle purity	±1%	90.02 (−3.23)%	75.60 (0.52)%
Reaction kinetic parameter	References		
Catalyst activity	[−100,100]	1.13	2.30
Metals activity	[0.5, 1.5]	1.01	1.45
Acid activity	[0.5, 1.5]	1.14	1.28
Aromatics selectivity	[0.5, 1.5]	1.11	1.65
Hydrogenolysis activity	[0.5, 1.5]	1.25	1.18
Catalyst coke capacity	[0.5, 1.5]	1.26	1.73
Coke yield penalty	[0.5, 1.5]	1.52	0.78
Coke hydrogen purity penalty	[0.5, 1.5]	1.10	1.60

Source: Khor (2019). Reproduced with permission of IntechOpen.

selectivity. The main input parameters are catalyst weight and reactor bed coke fraction. Table 9.1 summarizes the key variables that the application uses for process monitoring and their typical values.

9.1.5.2 Tuning Strategies

To tune the model to achieve the desired values and results, we balance between tuning the group of parameters for the reactor kinetics and that for the feed compositions. We can reduce the variation in one group by letting parameters move in the other group – the deliberation as to which to allow for more movement/adjustment depends on the feedback from the site or a process specialist on their relative importance. If the deliberation is not to change the feed composition much while letting the reactor tuning parameters move more, then use scaling factors of 1 for both parameter groups with appropriate lower and upper bounds for the parameters. (The solved parameters should not be at bounds.)

However, if the composition of the synthesized feed is significantly different from the actual plant condition, even allowing for the feed composition tuning parameters to move over a wide range may not help with convergence. Therefore, it is important to have a good starting feed composition, which can be achieved by calculating the feed distillation cutpoints at the same time at which the samples are taken.

Two other tuning strategies involve using calculated variables on the reformate yields and reactor total endotherms.

9.1.5.3 Reformate Yields

Reformate yield is given by the ratio of the reformate stream flow rate to that of the feed stream (typically on a mass basis). A calculated tag is generally not weighted and thus excluded from the objective function calculation. Instead, the tolerances for the two flow measurements used to calculate the yield (i.e. for the reformate and feed streams) are tightened. However, for the purpose of experimenting to determine ways to improve the model, we put a weight on the reformate yield tag and use a tight tolerance by specifying the standard deviation value to an artificially small number but one in which the model is still solvable in reasonable computational time. The resulting model solution gives information on the candidate independent variables whose bounds can be reasonably relaxed, measurement models whose tolerances (i.e. standard deviations) can be adjusted or relaxed, as well as possibly faulty or problematic instrumentation (due to malfunction or calibration issues). Note that this practice highlights a difference between using a data reconciliation application as compared to using a predictive optimization tool, in which the latter attempts to meet the reformate yields and other variables tightly.

9.1.5.4 Reactor Total Endotherms

Reactor total endotherms are the sum of the temperature differences (i.e. delta in temperatures) of each of the reforming reactors. This variable represents the temperature change and hence the reaction that takes place in a reactor, which can be the desirable endothermic reactions of dehydrogenation of naphthene to aromatic compounds and/or dehydrocyclization of paraffins to naphthenes or the undesirable (but necessary) exothermic reactions of hydrocracking and/or hydrocyclization. There are two relations available (which hence give rise to uncertainty) to calculate the temperature difference for a reactor: (i) take the difference between the temperature transmitters at the outlet of a preheater (i.e. upstream furnace) and that at the inlet of a furnace downstream; or (ii) take the difference between the reactor temperature indicators (i.e. thermocouples at the top skin and bottom skin). It is noted that when facing catalyst issues, the second relation may give a higher value than what is theoretically attainable due to heat loss that is actually experienced but not accounted for. Similar to the reformate yield being a calculated variable, it is also generally our practice to unweight (or to turn off) the total endotherm tag because it is an artificial or pseudo tag that may not physically exist at a site.

9.1.6 Concluding Remarks

In this work, we discuss a systematic workflow to develop a data reconciliation model for a catalytic reforming process to produce automotive fuels. We present the mathematical model formulation in the form of a nonlinear optimization model developed on a commercial software platform. The model is deployed as an online application to improve process advisory and monitoring at a refinery, besides providing a rich data source for a multitude of other associated relevant uses. We show the utility of such an application to contribute toward producing energy in an environmentally sustainable manner through an optimal process operation approach with reduced off-spec fuel products.

9.2 Industrial Case Study 2: Refinery Configuration for Low-Benzene Fuel Production

The ongoing global quest for clean fuel production has identified benzene as a gasoline component that should be reduced. The US Environmental Protection Agency limits benzene to less than 0.62 volume percent (vol%) on all US gasoline since 2011 under its Mobile Sources Air Toxics Phase 2 (MSAT II) regulations (United States Environmental Protection Agency (EPA) 2016). Similarly, gasoline for the European Union market is subject to Euro V (the current standard since 2009) and Euro VI fuel regulations that stipulate a maximum benzene amount of 1.0 vol% besides other requirements. The Malaysian Standards Euro 4M specifies a limit of 3.5 vol% benzene (Department of Environment Malaysia 2017).

This case study investigates the synthesis of a petroleum naphtha processing facility to produce low-benzene gasoline as a main product. We develop a conceptual design of a process or a sequence of processes to upgrade the naphtha feed so that it meets the MSAT II (or other equivalent) specifications for a product while attaining the highest possible octane number.

9.2.1 Problem Statement

The following data are assumed available:

- a set of refinery process units with known (fixed) yields, capacities, and stoichiometric coefficients for conversion operations;
- a set of inlet process streams with known compositions;
- a set of refined products with known market demands; costs of raw materials;
- prices of the saleable products;
- capital and operating costs to purchase, install, and operate the process units.

We wish to determine an optimal selection and sequencing of the units as well as their resulting interconnections and associated stream flow rates.

9.2.2 Superstructure Representation

A superstructure (STN-based) is developed for this case study, as shown in Figure 9.7. The major process units or technologies considered are categorized into three sections of a refinery to facilitate modeling: (i) front-end processing comprising hydrotreating, adsorption, and distillation; (ii) middle-end (intermediary) processing comprising two-phase separation, and gasoline blending; and (iii) back-end processing comprising naphtha splitting, catalytic reforming, hydrotreating, isomerization, alkylation, two-phase separation, distillation, and extractive distillation.

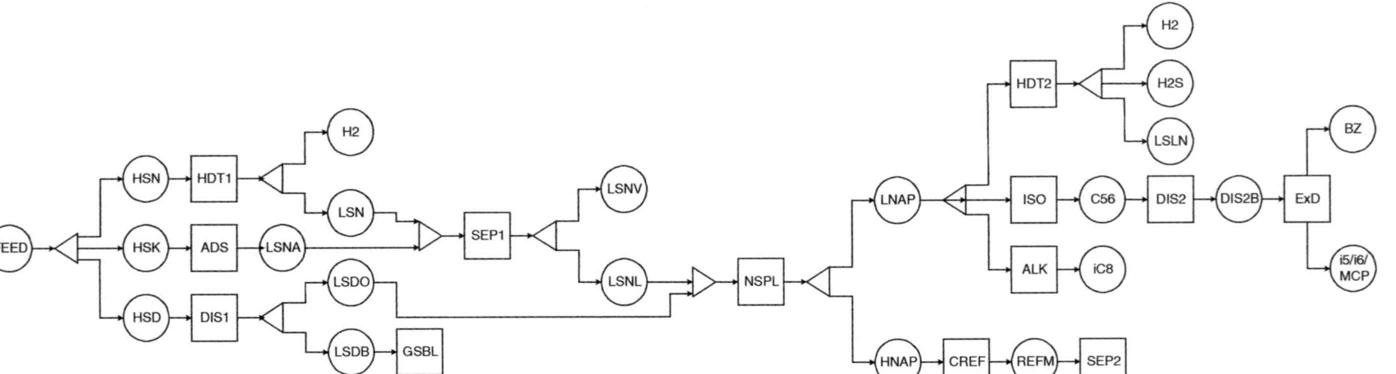

Figure 9.7 Superstructure representation (STN-based) for low-benzene production problem with the following legend on symbols used.

State		Task	
FEED	Feed stream from crude distillation unit	HDT1	Front-end hydrotreating
HSN	Heavy straight-run naphtha	ADS	Adsorption
HSD	Heavy straight-run diesel	DIS1	Front-end distillation
LSDO	Low sulfur distillate overhead	SEP1	Front-end two-phase separation
LSDO	Low sulfur distillate bottoms	NSPL	Naphtha splitting
		CREF	Catalytic reforming
		HDT2	Back-end hydrotreating
		ISO	Isomerization
		ALK	Alkylation
		DIS2	Back-end distillation
		ExD	Extractive distillation

9.2.3 Model Formulation

A general formulation on the disjunctive constraints (or disjunctions) for unit selection can be given by:

$$\begin{bmatrix} Y_k \\ \left.\begin{array}{l} \sum\limits_{j\in \text{FEED}} x_{i,j}^{\text{feed}} = \sum\limits_{j\in K_{\text{in}}} x_{i,j}^{\text{in}} \\ \sum\limits_{j\in K_{\text{in}}} x_{k,j}^{\text{in}} = \sum\limits_{j\in K_{\text{out}}} x_{k,j}^{\text{out}}, \quad \forall k \end{array}\right\} \begin{array}{c}\text{total}\\ \text{flow}\\ \text{balance}\end{array} \\ \left.\begin{array}{l} x_{k,j}^{\text{feed}} = x_{k,j}^{\text{in}}, \quad \forall j \\ x_{k,j}^{\text{in}} = x_{k,j}^{\text{out}}, \quad \forall j \end{array}\right\} \begin{array}{c}\text{component}\\ \text{balance}\end{array} \\ c_k = \gamma_k + \alpha_k F^{\beta} \end{bmatrix} \vee \begin{bmatrix} \neg Y_k \\ x = 0 \\ c = 0 \end{bmatrix} \tag{9.2}$$

Disjunction on selection of front-end hydrotreating operation:

$$\begin{bmatrix} Y_{\text{HDT1}} \\ \sum\limits_{j\in \text{FEED}} x_j^{\text{feed}} = \sum\limits_{j\in \text{FEED}} \left(x_{\text{HDT1},j}^{\text{in}} + x_{\text{ADS},j}^{\text{in}} + x_{\text{DIS1},j}^{\text{in}} \right) \\ x_j^{\text{feed}} = x_{\text{HDT1},j}^{\text{in}} + x_{\text{ADS},j}^{\text{in}} + x_{\text{DIS1},j}^{\text{in}}, \quad \forall j \in \text{FEED} \\ c_{\text{HDT1}} = \gamma_{\text{HDT1}} + \alpha_{\text{HDT1}} \sum\limits_{j\in \text{FEED}} x_j^{\text{feed}} \end{bmatrix} \vee \begin{bmatrix} \neg Y_{\text{HDT1}} \\ x_j^{\text{feed}} = 0 \\ x_{\text{HDT1},j}^{\text{in}} = 0 \\ c_{\text{HDT1}} = 0 \end{bmatrix} \tag{9.3}$$

Disjunction on selection of adsorption operation:

$$\begin{bmatrix} Y_{\text{ADS}} \\ \sum\limits_{j\in \text{FEED}} x_j^{\text{feed}} = \sum\limits_{j\in \text{FEED}} x_{\text{ADS},j}^{\text{in}} \\ x_j^{\text{feed}} = x_{\text{ADS},j}^{\text{in}}, \quad \forall j \in \text{FEED} \\ x_{\text{ADS},j}^{\text{in}} = x_{\text{ADS},j}^{\text{out}}, \quad j = \text{sulfur} \\ c_{\text{ADS}} = \gamma_{\text{ADS}} + \alpha_{\text{ADS}} \sum\limits_{j\in \text{FEED}} x_j^{\text{feed}} \end{bmatrix} \vee \begin{bmatrix} \neg Y_{\text{ADS}} \\ x_{\text{ADS},j}^{\text{in}} = 0 \\ c_{\text{ADS}} = 0 \end{bmatrix} \tag{9.4}$$

Disjunction on selection of front-end distillation operation:

$$\begin{bmatrix} Y_{\text{DIS1}} \\ \sum_{j\in\text{FEED}} x_j^{\text{feed}} = \sum_{j\in\text{FEED}} x_{\text{DIS1},j}^{\text{in}} \\ x_j^{\text{feed}} = x_{\text{DIS1},j}^{\text{in}}, \quad \forall j \\ \sum_{j\in\text{FEED}} x_{\text{DIS1},j}^{\text{in}} = \sum_{j\in\text{DIS1}_{\text{out}}^{\text{top}}} x_{\text{DIS1},j}^{\text{out,top}} + \sum_{j\in\text{DIS1}_{\text{out}}^{\text{btm}}} x_{\text{DIS1},j}^{\text{out,btm}} \\ x_{\text{DIS1},j}^{\text{in}} = x_{\text{DIS1},j}^{\text{out,top}} + x_{\text{DIS1},j}^{\text{out,btm}}, \forall j \in \text{FEED} \cap \text{DIS1}_{\text{out}}^{\text{top}} \cap \text{DIS1}_{\text{out}}^{\text{btm}} \\ \sum_{j\in\text{DIS1}_{\text{out}}^{\text{top}}} x_{\text{DIS1},j}^{\text{in}} = \sum_{j\in\text{DIS1}_{\text{out}}^{\text{top}}} x_{\text{DIS1},j}^{\text{out,top}} \\ x_{\text{DIS1},j}^{\text{in}} = x_{\text{DIS1},j}^{\text{out,top}}, \quad \forall j \in \text{DIS1}_{\text{out}}^{\text{top}} \\ \sum_{j\in\text{DIS1}_{\text{out}}^{\text{btm}}} x_{\text{DIS1},j}^{\text{in}} = \sum_{j\in\text{DIS1}_{\text{out}}^{\text{btm}}} x_{\text{DIS1},j}^{\text{out,btm}} \\ x_{\text{DIS1},j}^{\text{in}} = x_{\text{DIS1},j}^{\text{out,btm}}, \quad \forall j \in \text{DIS1}_{\text{out}}^{\text{btm}} \\ c_{\text{DIS1}} = \gamma_{\text{DIS1}} + \alpha_{\text{DIS1}} \sum_{j\in\text{FEED}} x_j^{\text{feed}} \end{bmatrix} \vee \begin{bmatrix} \neg Y_{\text{DIS1}} \\ x_{\text{DIS1},j}^{\text{in}} = 0 \\ x_{\text{DIS1},j}^{\text{out,top}} = 0 \\ x_{\text{DIS1},j}^{\text{out,btm}} = 0 \\ c_{\text{DIS1}} = 0 \end{bmatrix} \tag{9.5}$$

Disjunction on selection of front-end two-phase separator:

$$\begin{bmatrix} Y_{\text{SEP1}} \\ \sum_{j\in\text{HDT1}_{\text{out}}} x_{\text{HDT1},j}^{\text{out}} + \sum_{j\in\text{ADS}_{\text{out}}} x_{\text{ADS},j}^{\text{out}} = \sum_{j\in\text{SEP1}_{\text{in}}} x_{\text{SEP1},j}^{\text{in}} \\ x_{\text{HDT1},j}^{\text{out}} + x_{\text{ADS},j}^{\text{out}} = x_{\text{SEP1},j}^{\text{in}}, \quad \forall j \in \text{HDT1}_{\text{out}} \cap \text{ADS}_{\text{out}} \cap \text{SEP1}_{\text{in}} \\ \sum_{j\in\text{SEP1}_{\text{out}}^{\text{top}}} x_{\text{SEP1},j}^{\text{in}} = \sum_{j\in\text{SEP1}_{\text{out}}^{\text{top}}} x_{\text{SEP1},j}^{\text{out,top}} \\ x_{\text{SEP1},j}^{\text{in}} = x_{\text{SEP1},j}^{\text{out,top}}, \quad \forall j \in \text{SEP1}_{\text{out}}^{\text{top}} \\ \sum_{j\in\text{SEP1}_{\text{out}}^{\text{btm}}} x_{\text{SEP1},j}^{\text{in}} = \sum_{j\in\text{SEP1}_{\text{out}}^{\text{btm}}} x_{\text{SEP1},j}^{\text{out,btm}} \\ x_{\text{SEP1},j}^{\text{in}} = x_{\text{SEP1},j}^{\text{out,btm}}, \quad \forall j \in \text{SEP1}_{\text{out}}^{\text{btm}} \\ c_{\text{SEP1}} = \gamma_{\text{SEP1}} + \alpha_{\text{SEP1}} \left(\sum_{j\in\text{HDT}_{\text{out}}} x_{\text{HDT1},j}^{\text{out}} + \sum_{j\in\text{ADS}_{\text{out}}} x_{\text{ADS},j}^{\text{out}} \right) \end{bmatrix} \vee \begin{bmatrix} \neg Y_{\text{SEP1}} \\ x_{\text{SEP1},j}^{\text{in}} = 0 \\ x_{\text{SEP1},j}^{\text{out,top}} = 0 \\ x_{\text{SEP1},j}^{\text{out,btm}} = 0 \\ c_{\text{SEP1}} = 0 \end{bmatrix} \tag{9.6}$$

where $x_{\text{SEP1},j}^{\text{out,top}}$ = outlet flow rate of the top product for component j from the front-end two-phase separator,

Disjunction on selection of gasoline blending operation:

$$\begin{bmatrix} Y_{\text{GSBL}} \\ \sum_{j\in\text{DIS1}_{\text{out}}^{\text{btm}}} x_{\text{DIS1},j}^{\text{out,btm}} = \sum_{j\in\text{GSBL}_{\text{in}}} x_{\text{GSBL},j}^{\text{in}} \\ c_{\text{GSBL}} = \gamma_{\text{GSBL}} + \alpha_{\text{GSBL}} \sum_{j\in\text{DIS1}_{\text{out}}^{\text{btm}}} x_{\text{DIS1},j}^{\text{out,btm}} \end{bmatrix} \vee \begin{bmatrix} \neg Y_{\text{GSBL}} \\ x_{\text{GSBL},j}^{\text{in}} = 0 \\ c_{\text{GSBL}} = 0 \end{bmatrix} \tag{9.7}$$

Disjunction on selection of naphtha splitting operation:

$$\begin{bmatrix} Y_{\text{NSPL}} \\ \sum_{j\in \text{SEP1}_{\text{out}}^{\text{btm}}} x_{\text{SEP1},j}^{\text{out,btm}} + \sum_{j\in \text{DIS1}_{\text{out}}^{\text{top}}} x_{\text{DIS1},j}^{\text{out,top}} = \sum_{j\in \text{NSPL}_{\text{in}}} x_{\text{NSPL},j}^{\text{in}} \\ x_{\text{SEP1},j}^{\text{out,btm}} + x_{\text{DIS1},j}^{\text{out,top}} = x_{\text{NSPL},j}^{\text{in}}, \quad \forall j \\ \sum_{j\in \text{NSPL}_{\text{out}}^{\text{top}}} x_{\text{NSPL},j}^{\text{in}} = \sum_{j\in \text{NSPL}_{\text{out}}^{\text{top}}} x_{\text{NSPL},j}^{\text{out,top}} \\ x_{\text{NSPL},j}^{\text{in}} = x_{\text{NSPL},j}^{\text{out,top}}, \quad \forall j \in \text{NSPL}_{\text{out}}^{\text{top}} \\ \sum_{j\in \text{NSPL}_{\text{out}}^{\text{btm}}} x_{\text{NSPL},j}^{\text{in}} = \sum_{j\in \text{NSPL}_{\text{out}}^{\text{btm}}} x_{\text{NSPL},j}^{\text{out,btm}} \\ x_{\text{NSPL},j}^{\text{in}} = x_{\text{NSPL},j}^{\text{out,btm}}, \quad \forall j \in \text{NSPL}_{\text{out}}^{\text{btm}} \\ c_{\text{NSPL}} = \gamma_{\text{NSPL}} + \alpha_{\text{NSPL}} \left(\sum_{j\in \text{SEP1}_{\text{out}}^{\text{btm}}} x_{\text{SEP1},j}^{\text{out,btm}} + \sum_{j\in \text{DIS1}_{\text{out}}^{\text{top}}} x_{\text{DIS1},j}^{\text{out,top}} \right) \end{bmatrix} \vee \begin{bmatrix} \neg Y_{\text{NSPL}} \\ x_{\text{NSPL},j}^{\text{in}} = 0, \quad \forall j \\ x_{\text{NSPL},j}^{\text{out,top}} = 0, \quad \forall j \\ x_{\text{NSPL},j}^{\text{out,btm}} = 0, \quad \forall j \\ c_{\text{NSPL}} = 0 \end{bmatrix} \tag{9.8}$$

Disjunction on selection of catalytic reforming operation:

$$\begin{bmatrix} Y_{\text{CREF}} \\ \sum_{j\in \text{NSPL}_{\text{out}}^{\text{btm}}} x_{\text{NSPL},j}^{\text{out,btm}} = \sum_{j\in \text{CREF}_{\text{in}}} x_{\text{CREF},j}^{\text{in}} \\ x_{\text{NSPL},j}^{\text{out,btm}} = x_{\text{CREF},j}^{\text{in}}, \quad \forall j \in \text{NSPL}_{\text{out}}^{\text{btm}} \cap \text{CREF}_{\text{in}} \\ \sum_{j\in \text{CREF}_{\text{in}}} x_{\text{CREF},j}^{\text{in}} = \sum_{j\in \text{CREF}_{\text{out}}} x_{\text{CREF},j}^{\text{out}} \\ x_{\text{CREF},j}^{\text{out}} = y_{\text{CREF},j} x_{\text{CREF},j}^{\text{in}}, \quad \forall j \in \text{CREF}_{\text{in}} \cap \text{CREF}_{\text{out}} \\ c_{\text{CREF}} = \gamma_{\text{CREF}} + \alpha_{\text{CREF}} \sum_{j\in \text{NSPL}_{\text{out}}^{\text{btm}}} x_{\text{NSPL},j}^{\text{out,btm}} \end{bmatrix} \vee \begin{bmatrix} \neg Y_{\text{CREF}} \\ x_{\text{CREF},j}^{\text{in}} = 0, \quad \forall j \\ x_{\text{CREF},j}^{\text{out}} = 0, \quad \forall j \\ c_{\text{CREF}} = 0 \end{bmatrix} \tag{9.9}$$

where $y_{\text{CREF},j}$ = yield of reformer for component j.

Disjunction on selection of back-end hydrotreater:

$$\begin{bmatrix} Y_{\text{HDT2}} \\ \sum_{j\in \text{NSPL}_{\text{out}}^{\text{top}}} x_{\text{NSPL},j}^{\text{out,top}} = \sum_{j\in \text{HDT2}_{\text{in}}} x_{\text{HDT2},j}^{\text{in}} \\ x_{\text{NSPL},j}^{\text{out,top}} = x_{\text{HDT2},j}^{\text{in}}, \quad \forall j \in \text{NSPL}_{\text{out}}^{\text{top}} \cap \text{HDT2}_{\text{in}} \\ \sum_{j\in \text{HDT2}_{\text{in}}} x_{\text{HDT2},j}^{\text{in}} = \sum_{j\in \text{HDT2}_{\text{out}}} x_{\text{HDT2},j}^{\text{out}} + \\ c_{\text{HDT2}} = \gamma_{\text{HDT2}} + \alpha_{\text{HDT2}} \sum_{j\in \text{NSPL}_{\text{out}}^{\text{btm}}} x_{\text{NSPL},j}^{\text{out,btmp}} \end{bmatrix} \vee \begin{bmatrix} \neg Y_{\text{HDT2}} \\ x_{\text{HDT2},j}^{\text{in}} = 0 \\ c_{\text{HDT2}} = 0 \end{bmatrix} \tag{9.10}$$

Disjunction on selection of isomerization:

$$\begin{bmatrix} Y_{\text{ISO}} \\ \sum_{j\in \text{NSPL}_{\text{out}}^{\text{top}}} x_{\text{NSPL},j}^{\text{out,top}} = \sum_{j\in \text{ISO}_{\text{in}}} x_{\text{ISO},j}^{\text{in}} \\ x_{\text{NSPL},j}^{\text{out,top}} = x_{\text{ISO},j}^{\text{in}}, \quad \forall j \in \text{NSPL}_{\text{out}}^{\text{top}} \cap \text{ISO}_{\text{in}} \\ \sum_{j\in \text{ISO}_{\text{in}}} x_{\text{ISO},j}^{\text{in}} = \sum_{j\in \text{ISO}_{\text{out}}} x_{\text{ISO},j}^{\text{out}} \\ c_{\text{ISO}} = \gamma_{\text{ISO}} + \alpha_{\text{ISO}} \sum_{j\in \text{NSPL}_{\text{out}}^{\text{top}}} x_{\text{NSPL},j}^{\text{out,top}} \end{bmatrix} \vee \begin{bmatrix} \neg Y_{\text{ISO}} \\ x_{\text{ISO},j}^{\text{in}} = 0 \\ x_{\text{ISO},j}^{\text{out}} = 0 \\ c_{\text{ISO}} = 0 \end{bmatrix} \tag{9.11}$$

Disjunction on selection of alkylation:

$$\begin{bmatrix} Y_{\text{ALK}} \\ \sum_{j\in \text{NSPL}_{\text{out}}^{\text{top}}} x_{\text{NSPL},j}^{\text{out,top}} = \sum_{j\in \text{ALK}_{\text{in}}} x_{\text{ALK},j}^{\text{in}} \\ x_{\text{NSPL},j}^{\text{out,top}} = x_{\text{ALK},j}^{\text{in}}, \quad \forall j \in \text{NSPL}_{\text{out}}^{\text{top}} \cap \text{ALK}_{\text{in}} \\ c_{\text{ALK}} = \gamma_{\text{ALK}} + \alpha_{\text{ALK}} \sum_{j\in \text{NSPL}_{\text{out}}^{\text{top}}} x_{\text{NSPL},j}^{\text{out,top}} \end{bmatrix} \vee \begin{bmatrix} \neg Y_{\text{ALK}} \\ x_{\text{ALK},j}^{\text{in}} = 0 \\ c_{\text{ALK}} = 0 \end{bmatrix} \tag{9.12}$$

Disjunction on selection of back-end two-phase separator:

$$\begin{bmatrix} Y_{\text{SEP2}} \\ \sum_{j\in \text{CREF}_{\text{out}}} x_{\text{CREF},j}^{\text{out}} = \sum_{j\in \text{SEP2}_{\text{out}}^{\text{top}}} x_{\text{SEP2},j}^{\text{out,top}} + \sum_{j\in \text{SEP2}_{\text{out}}^{\text{btm}}} x_{\text{SEP2},j}^{\text{out,btm}} \\ x_{\text{CREF},j}^{\text{out}} = x_{\text{SEP2},j}^{\text{out,top}} + x_{\text{SEP2},j}^{\text{out,btm}}, \quad \forall j \in \text{CREF}_{\text{out}} \cap \left(\text{SEP2}_{\text{out}}^{\text{top}} \cup \text{SEP2}_{\text{out}}^{\text{btm}}\right) \\ c_{\text{SEP2}} = \gamma_{\text{SEP2}} + \alpha_{\text{SEP2}} \sum_{j\in \text{CREF}_{\text{out}}} x_{\text{CREF},j}^{\text{out}} \end{bmatrix} \vee \begin{bmatrix} \neg Y_{\text{SEP2}} \\ x_{\text{CREF},j}^{\text{out}} = 0 \\ x_{\text{SEP2},j}^{\text{out,top}} = 0 \\ x_{\text{SEP2},j}^{\text{out,btm}} = 0 \\ c_{\text{SEP2}} = 0 \end{bmatrix} \tag{9.13}$$

Disjunction on selection of back-end distillation operation:

$$\begin{bmatrix} Y_{\text{DIS2}} \\ \sum_{j\in \text{ISO}_{\text{out}}} x_{\text{ISO},j}^{\text{out}} = \sum_{j\in \text{DIS2}_{\text{out}}^{\text{top}}} x_{\text{DIS2},j}^{\text{out,top}} + \sum_{j\in \text{DIS2}_{\text{out}}^{\text{btm}}} x_{\text{DIS2},j}^{\text{out,btm}} \\ x_{\text{ISO},j}^{\text{out}} = x_{\text{DIS2},j}^{\text{out,top}} + x_{\text{DIS2},j}^{\text{out,btm}}, \quad \forall j \in \text{ISO}_{\text{out}} \cap \left(\text{DIS2}_{\text{out}}^{\text{top}} \cup \text{DIS2}_{\text{out}}^{\text{btm}} s\right) \\ c_{\text{DIS2}} = \gamma_{\text{DIS2}} + \alpha_{\text{DIS2}} \sum_{j\in \text{ISO}_{\text{out}}} x_{\text{ISO},j}^{\text{out}} \end{bmatrix} \vee \begin{bmatrix} \neg Y_{\text{DIS2}} \\ x_{\text{ISO},j}^{\text{out}} = 0 \\ x_{\text{DIS2},j}^{\text{out,top}} = 0 \\ x_{\text{DIS2},j}^{\text{out,btm}} = 0 \\ c_{\text{DIS2}} = 0 \end{bmatrix} \tag{9.14}$$

Disjunction on selection of extractive distillation operation:

$$\begin{bmatrix} Y_{\text{ED}} \\ \sum_{j \in \text{DIS2}_{\text{out}}^{\text{btm}}} x_{\text{DIS2},j}^{\text{out,btm}} = \sum_{j \in \text{ED}_{\text{in}}} x_{\text{ED},j}^{\text{in}} \\ x_{\text{DIS2},j}^{\text{out,btm}} = x_{\text{ED},j}^{\text{in}}, \quad \forall j \in \text{DIS2}_{\text{out}}^{\text{btm}} \cap \text{ED}_{\text{in}} \\ c_{\text{ED}} = \gamma_{\text{ED}} + \alpha_{\text{ED}} \sum_{j \in \text{DIS2}_{\text{out}}^{\text{btm}}} x_{\text{DIS2},j}^{\text{out,btm}} \end{bmatrix} \vee \begin{bmatrix} \neg Y_{\text{ED}} \\ x_{\text{DIS2},j}^{\text{out,btm}} = 0 \\ x_{\text{ED},j}^{\text{in}} = 0 \\ c_{\text{ED}} = 0 \end{bmatrix} \tag{9.15}$$

Logical constraints:

$$Y_{\text{HDT1}} \Rightarrow \neg Y_{\text{ADS}} \tag{9.16}$$

$$Y_{\text{ADS}} \Rightarrow \neg Y_{\text{HDT1}} \tag{9.17}$$

$$Y_{\text{HDT1}} \wedge \neg Y_{\text{ADS}} \Rightarrow Y_{\text{DIS1}} \tag{9.18}$$

$$Y_{\text{SEP1}} \Rightarrow Y_{\text{HDT1}} \vee Y_{\text{ADS}} \vee Y_{\text{DIS1}} \tag{9.19}$$

$$Y_{\text{SEP1}} \vee Y_{\text{DIS1}} \Rightarrow Y_{\text{NSPL}} \tag{9.20}$$

$$Y_{\text{DIS1}} \Rightarrow Y_{\text{NSPL}} \vee Y_{\text{GSBL}} \tag{9.21}$$

$$Y_{\text{SEP1}} \Rightarrow Y_{\text{NSPL}} \tag{9.22}$$

$$Y_{\text{GSBL}} \Rightarrow Y_{\text{DIS1}} \tag{9.23}$$

$$Y_{\text{NSPL}} \Rightarrow Y_{\text{HDT2}} \vee Y_{\text{ISO}} \vee Y_{\text{CREF}} \vee Y_{\text{ALK}} \tag{9.24}$$

$$Y_{\text{HDT2}} \Rightarrow Y_{\text{NSPL}} \tag{9.25}$$

$$Y_{\text{ISO}} \Rightarrow Y_{\text{NSPL}} \tag{9.26}$$

$$Y_{\text{CREF}} \Rightarrow Y_{\text{NSPL}} \tag{9.27}$$

$$Y_{\text{ALK}} \Rightarrow Y_{\text{NSPL}} \tag{9.28}$$

$$Y_{\text{CREF}} \Rightarrow Y_{\text{SEP2}} \tag{9.29}$$

$$Y_{\text{DIS2}} \Rightarrow Y_{\text{ISO}} \tag{9.30}$$

$$Y_{\text{ED}} \Rightarrow Y_{\text{DIS2}} \tag{9.31}$$

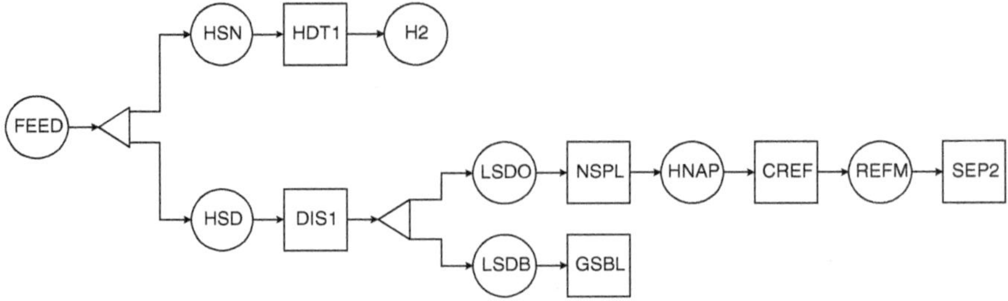

Figure 9.8 Optimal refinery configuration for low-benzene fuel production.

9.2.4 Preliminary Computational Results

An optimal refinery configuration for low-benzene fuel production is shown in Figure 9.8. The main processing route involves catalytically reforming (CREF unit) a heavy naphtha (HNAP) stream from the outlet (which is the bottom stream) of a naphtha splitter (NSPL unit). The NSPL feed is an overhead stream (with low sulfur content) from a distillation column that processes a high-sulfur diesel fraction of the feedstock material.

9.3 Chapter Summary

This chapter presents several applications of the foregoing optimization modeling framework introduced in earlier chapters. Four case studies on configuration design illustrate problems of industrial size and significance for the petroleum refining industry. Each of the applied examples emphasizes different aspects of importance to this process synthesis problem type.

References

Amand, T., Heyen, G., and Kalitventzeff, B. (2001). Plant monitoring and fault detection: Synergy between data reconciliation and principal component analysis. *Computers & Chemical Engineering* 25 (4): 501–507.

Britt, H.I. and Luecke, R.H. (1973). The estimation of parameters in nonlinear, implicit models. *Technometrics* 15 (2): 233–247.

Crowe, C.M. (1989). Observability and redundancy of process data for steady state reconciliation. *Chemical Engineering Science* 44 (12): 2909–2917.

Department of Environment Malaysia. 2017. *Application Guidelines for Motor Vehicle Exhaust Gas and Noise Emission Type Approval.* (ed. E. M. F. Standard), 24.

Gill, P.E., Murray, W., and Wright, M.H. (1981). *Practical Optimization*, 133. London, UK: Academic Press.

Heyen, G. and Kalitventzeff, B. (2006). Process monitoring and data reconciliation. In: *Computer Aided Process and Product Engineering* (ed. L. Puigjaner and G. Heyen). Weinheim, Germany: Wiley-VCH Verlag GmbH.

Khor, C. (2019). High-octane gasoline production from catalytic naphtha reforming. In: *Petroleum Chemicals* (ed. M. Zoveidavianpoor). London, UK: InTechOpen.

Lid, T. and Skogestad, S. (2008). Data reconciliation and optimal operation of a catalytic naphtha reformer. *Journal of Process Control* 18 (3): 320–331.

Moser, M. and Sadler, C.C. (2002). Reforming—Industrial. In: *Encyclopedia of Catalysis*. Wiley.

Moser, M.D. and Bogdan, P.L. (2008). Catalytic reforming. In: *Handbook of Heterogeneous Catalysis*. Wiley-VCH Verlag GmbH & Co. KGaA.

Rahimpour, M.R., Jafari, M., and Iranshahi, D. (2013). Progress in catalytic naphtha reforming process: a review. *Applied Energy* 109: 79–93.

Schneider Electric (2017). *ROMeo Process Optimization Solution: Rigorous Online Modeling with Equation-based Optimization*. http://software.schneider-electric.com/pdf/datasheet/romeo-process-optimization/ (accessed 11 November 2021).

Sinfelt, J.H. (1981). Catalytic reforming of hydrocarbons. In: *Catalysis – Science and Technology* (ed. J.R. Anderson and M. Boudart), 258–298. Berlin Heidelberg: Springer-Verlag.

Stanley, G.M. and Mah, R.S.H. (1981). Observability and redundancy in process data estimation. *Chemical Engineering Science* 36 (2): 259–272.

United.States Environmental Protection Agency (EPA) (2020). *Gasoline Mobile Source Air Toxics*. EPA 2016 [cited October 19 2020]. https://www.epa.gov/gasoline-standards/gasoline-mobile-source-air-toxics (accessed 11 November 2021).

Wang, D. and Romagnoli, J.A. (2005). Generalized T distribution and its applications to process data reconciliation and process monitoring. *Transactions of the Institute of Measurement and Control* 27 (5): 367–390.

Wu, S.X., Ye, Q., Chen, C., and Gun, X.S. (2016). Research on data reconciliation based on generalized T distribution with historical data. *Neurocomputing* 175: 808–815.

Summary and Conclusions

The book applies a superstructure optimization approach to refinery design that serves to embed many if not all feasible processing alternatives. We develop logical constraints on design and structural specifications for the alternatives and incorporate them within mixed integer linear programming (MILP) and generalized disjunctive programming (GDP) model formulations. The former is solved by considering binary variables for discrete decisions directly, while the latter is handled by using disjunctions in arriving at an optimal topology or configuration that generally agrees with typical industrial setup as shown through the examples and case studies discussed. Scope exists for incorporating other practical specifications on qualitative heuristics, knowledge, and experience pertaining to refinery design in a suitable modeling framework.

Model-Based Optimization for Petroleum Refinery Configuration Design, First Edition. Cheng Seong Khor.

Index